THE UNIVERSAL MACHINE

*From the Dawn of Computing
to Digital Consciousness*

Ian Watson

THE UNIVERSAL MACHINE

*From the Dawn of Computing
to Digital Consciousness*

Copernicus Books

An Imprint of Springer Science + Business Media

© Springer-Verlag Berlin Heidelberg 2012

Published in the United States by Copernicus Books,
an imprint of Springer Science+Business Media.

Copernicus Books
Springer Science+Business Media
233 Spring Street
New York, NY 10013
www.springer.com

Library of Congress Control Number: 9783642281013

Manufactured in the United States of America.
Printed on acid-free paper

ISBN 978-3-642-28101-3   e-ISBN 978-3-642-28102-0

This book is dedicated to the memory of my grandfather, an airman, who gave his life in World War II, and to Alan Turing and all the men and women who worked alongside him, in secrecy and without any recognition at Bletchley Park to ensure that we all enjoy freedom.

Preface

Whilst working on this book several people asked me, "*What gave you the idea to write the book?*" Good question – the idea of writing a popular science book had been in my mind for a while. I like popular science documentaries and books; I think they help people better understand the world they live in. I also like histories and again it's important that people recognize the remarkable achievements of those who preceded them. I've read and enjoyed books like *Salt: A World History* and *Spice: The History of Temptation,* so at some point it must have come to me that a combination of popular science and history would be a good idea.

To write a popular history of the computer was a natural progression since I'm a computer scientist and have always had an interest in its history. But, to write a popular book on computing and leave out any speculation about what the future of artificial intelligence and robotics holds for us seemed daft. I had to include the exciting fun stuff! So there you have it, the idea for a book on the history and future of computing was conceived.

I then did some research; surely there must be other books just like this already? As I trawled through Amazon.com and my excellent local library looking for popular science and history books I had a pleasant surprise. There were lots of books about specific aspects of this story: Silicon Valley in the 60s and 70s, Bill Gates and the software entrepreneurs, biographies of people like Steve Jobs, detailed accounts of WWII code-breaking, books on early business computers, and books by futurologists imagining wonderful, terrifying or fanciful futures. But, there seemed to be no book that told the entire story from the first computers, through to the present day, plus a projection into the future. I had my idea and it seemed a gap in the market.

I've been working on this for about 3 years now. First doing the background research and deciding from the mountain of material what to include. I had the great luxury of being on sabbatical from *The University of Auckland* for the last year during which time I finished the book. I therefore must thank my university and the *Department of Computer Science*; it really was a joy and a privilege to be able to focus entirely on one project for a year. Two people in particular deserve thanks for reading and commenting on drafts of the chapters as I was writing them: Bob Doran, who is a professor in my department and curator of our excellent computer history displays,[1] and Duncan Campbell, a good friend and professional writer. Brian Carpenter, an Internet expert, commented on the Internet chapter and Dean Carter, a security expert, commented on the chapter on hacking. Several other friends read various drafts as the book progressed and author Nassau Hedron helped proof read the final version – I thank you all. Any errors or omissions are entirely of my own making and I encourage readers to leave any comments you have on the book's blog[2] – together we can improve the second edition.

I would like to thank my editor at Springer, Angela Lahee, for taking a chance on me, and her whole team for doing such an excellent job in producing the book. I must thank my friends who were very gracious when one of their number

---

[1] http://www.cs.auckland.ac.nz/historydisplays/
[2] www.universal-machine.com

appeared to have an entire year off! Perhaps this book will prove I *really* was working. Finally, I must thank my beautiful and patient wife who now knows far more about the history of computing than she ever wanted to.

Contents

Chapter 1

INTRODUCTION   1

Chapter 2

THE DAWN OF COMPUTING   9

*The Difference Engine; The First Computers;
The Grandfather of Computing; Babbage Goes Public;
The Analytical Engine; The Enchantress of Numbers;
The Difference Engine No.2; A Swedish Doppelgänger;
Babbage's Legacy*

Chapter 3

MARVELOUS MACHINES   41

*We the People...; Soldiers and Secretaries; The Sound of
Money; Weighing, Counting, Timing and Tabulating*

Chapter 4

COMPUTERS GO TO WAR   51

*The Turing Machine; Total War; Chasing an Enigma; Hitler's
Secret Writer; Colossus; Calculating Space; The Turing
Test; To Die Perchance to Sleep...; Turing's Legacy*

Contents

Chapter 5

COMPUTERS AND BIG BUSINESS   89

*ENIAC; EDVAC; The Mainframe; The Lyons Electronic
Office; Electronic Recording Machine-Accounting;
Fly the American Way*

Chapter 6

DEADHEADS AND PROPELLER HEADS   105

*Augmenting Human Intellect; Stanford University
and Silicon Valley; SAIL; Playing in the PARC;
Inventing the Future*

Chapter 7

THE COMPUTER GETS PERSONAL   125

*Woz; Phreaking, Jokes and Jobs; Homebrew and the Birth
of Apple; A PC on Every Desk; Killer Apps; It's All About the
Software; Pirates and Entrepreneurs; The Deal of the
Century; 1984*

Chapter 8

WEAVING THE WEB   161

*Vague But Exciting; Explosive Growth; A Network of
Networks; You've Got Mail!; An Electronic Notice Board;
The Browser Wars; A Needle in a Million Haystacks*

Chapter 9

DOTCOM   183

*Power to the People; Search and You Will Find; Everything
from A to Z; I Collect Broken Laser Pointers!; It's Good
to Share; The Dotcom Bubble*

Contents

Chapter 10

THE SECOND COMING 201

Stay Hungry, Stay Foolish; Change the World; What's NeXT; Some Light Entertainment; Act II; All Your Music in Your Pocket; I Wanted One so Much it Hurt?; Post-PC; RIP

Chapter 11

WEB 2.0 235

Let Me Share with You; Harvard's Face Books; Identical Twins; The Next Big Thing; Ambient Intimacy; LOLcats; Citizen Journalism; 140 Characters or Less; All the World's Knowledge

Chapter 12

DIGITAL UNDERWORLD 259

The Birth of Hacking; Captain Crunch; Cyberpunks; The Hacker Manifesto; Worms, Viruses and Trojans; Good Guys and Bad Guys; Hacktivism; Cyberwar

Chapter 13

MACHINES OF LOVING GRACE 285

How Far Is the Future?; Moore's Law; The Near Future; Ubiquitous Computing; Cloud Computing; Computing as a Utility; Hands-Free; Augmented Reality; A Grand Challenge

Chapter 14

DIGITAL CONSCIOUSNESS 307

The Imitation Game; The Chinese Room; The Robots Are Coming; Governing Lethal Behavior in Autonomous Robots; Resistance is Futile; Spooky Action; Reverse Engineering the Brain; Epilogue

Contents

APPENDIX I   331

APPENDIX II   335

FURTHER READING   343

Chapter 1

INTRODUCTION

Hi, you probably don't know me, but assuming you stick with this book then we're going to be spending quite a bit of time together. I'm sitting at my desk in my study in Auckland, New Zealand, writing this and you are reading these words. I have no idea where you are; you could be at home, on a bus, train or plane, in a hospital or perhaps sitting on a sunny beach somewhere. You may be reading this as a printed book, much as they've looked for centuries, or you may be reading it on a Kindle or iPad. Now here's the thing—had I written a book 50 years ago, I would have written the book with a typewriter or by hand with a pen. The book would have been printed and bound and you, or your ancestor, would have held and read the book and turned its pages.

In 2012 though, you can be fairly certain that I wrote this book using a computer, and I can be reasonably certain that many of you are reading this using an e-book reader. Something remarkable has happened in the last 50 or so years that has transformed the way I and every other author writes, and more recently, the way many people read. The computer has caused that change.

I'm writing this on my MacBook Pro using Google Docs in the cloud. I listen to music when I write; you may listen to music when you read. Ten years ago we would have used a hi-fi or perhaps a Walkman—now we're listening to music from our iTunes libraries from a computer or iPod. Once again computers have changed habits that had lasted for decades, or at least since the invention of the gramophone and radio. Think about this: in 1940 nobody had seen or used a computer. For millennia the world had got along just fine; children had been born, work had been done, wars fought and nobody had thought, "*I wish I*

I. Watson, *The Universal Machine,*
DOI 10.1007/978-3-642-28102-0_1,
© Springer-Verlag Berlin Heidelberg 2012

could have done that with a computer." Then at the end of World War II computers were developed, and a few years later the first computers were doing the payroll for a café company and selling airline tickets. A few years after that a bunch of techno-hippies around San Francisco started building their own personal computers, and by the end of the twentieth century the world had changed totally. Now babies can't be born, work isn't done and wars can't be fought without computers being involved.

How did this happen? Think of other iconic inventions of the last century—the Jumbo Jet for instance. The Boeing 747 first flew commercially in 1970; nearly 40 years later it is still in regular use. The motorcar really hasn't changed much in a 100 years. Sure they are now more comfortable, reliable, safer and faster, but back in the 1920s a car could carry four or more people and cruise all day at 50 miles an hour. Their basic technology hasn't changed much at all; they even had electric cars in the 1920s. Now look at your computer. Twenty years ago you probably didn't own a computer and you didn't need one for your daily work. Thirty years ago, apart from some science-fiction "*electronic brain*" you'd never seen one, unless you were a computer science student. Seventy years ago nobody could have seen one, because they didn't yet exist.

So in less than a single life span the computer has transformed almost everything in our society. John Lennon once sang, "*Imagine no possessions, I wonder if you can?*" Can you imagine a world without computers?

This book is a history book. It will take you back on a relatively short journey to the dawn of the computer and introduce you to many of the brilliant and interesting people who were influential in its development. Along the way, we'll make little diversions and detours to look at some of the wonderful things that computers can do, and finally we'll stop and look into the future and wonder at what may be just over the horizon. Of one thing we can be sure—we will certainly be amazed at what lies ahead.

Don't worry, this book wont be full of math, logic and incomprehensible formulae. It is not my aim to make you totally understand how a computer actually works. Indeed, even as a professor of computer science I don't really care much what goes on inside the box. Computers to me, and probably you, are a bit like cars—I just want to get in, turn the key and drive. I don't need to know the details of how the engine works. However, by reading a book about Formula One, you might at the end, have a better understanding of motor racing technology. I know that by the end of this book you'll have a better insight into how computers work and what makes them so uniquely useful and powerful.

Most books on the history of computers start with the abacus or other primitive counting devices. I'm not interested in abacuses because they are not "*universal machines.*" An abacus does one thing: simple arithmetic. You can't use an abacus to write a note or play a piece of music. It does one thing and one thing only.

I intend to show that computers are unlike other machines that people make and use, in that they are *universal.* They can be used to do many different tasks. Think of a DIY enthusiast's toolbox; it contains many tools, each with a specific purpose: a screwdriver, a hammer, a file, a wrench. Each has a limited range of tasks for which it can be used. The Swiss Army knife can do many things, but it's really just a collection of individual tools ingeniously combined to be convenient and pocketable.

Before we made machines people used tools; the earliest example of a universal tool must be fire. Our ancestors learnt to use this natural phenomenon to accomplish a wide range of tasks. It was used as a source of heat and light, and of course to cook food. Fire was also used to clear land of dense forest to create clearings where new grass could grow, fertilized by the wood ash. Grazing animals, like deer, would be drawn to these clearings and were more easily hunted than in thick forest. Fire could also be used to sharpen spear points and fire clay pots. It was harnessed to smelt metals so all manner of tools could be made. Finally, fire was even used for long-distance communication, by burning beacons at night and making smoke signals during the day.

Fire, of course, is not an artifact but rather something that early humans used to transform their environment. Perhaps the earliest example of an artifact, a real tool created by people, is the stone hand axe from Paleolithic times.

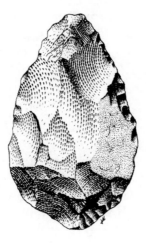

A stone hand axe

Stone hand axes have been found in Africa, Europe, and Asia and are mostly made from flint by striking suitable flints with a stone hammer (another stone) and in later examples with antler. The basic shape is a teardrop with cutting edges down the sides and point, with the handle being the fat end. If you are wondering how stone can be used to cut anything, you should remember that the edge of a piece of flint is as sharp as broken glass. Although the cutting edge dulls quite quickly, the edges can be easily re-sharpened by a process called *knapping*.

Archaeologists have shown, by demonstration, that stone axes can be used to: chop wood, de-bark trees and branches, sharpen spears for hunting, skin animals, butcher carcasses, scrape skins for leather-making, crack bones for marrow (a very high protein food source) and for a host of other tasks. Indeed, for Stone Age man the hand axe was his universal tool, and some archaeologists argue that their society couldn't function without these axes. They are what define them. They are what increased their productivity so that a society could develop in the downtime the surplus supported. The down-time allowed language, customs, religion and art to develop. New ideas also started to develop, like domesticating animals and planting crops. The hand axe may have been the precursor to civilization.

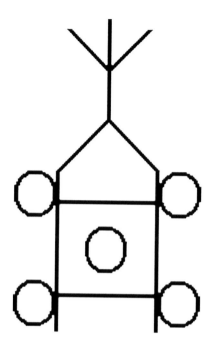

An image from the Bronocice pot, depicting a wheeled wagon
circa 3,400 BC

Another example of a universal tool is the wheel. There is
archaeological evidence of wheeled vehicles (images of
wagons and chariots on pottery) from around 4,000 BC across
Central Europe, the Caucasus and Mesopotamia (modern
Iraq). Nobody knows where or when the wheel was first
invented, but it seems the invention spread through the
civilizations around the Mediterranean very rapidly. It is not
the use of the wheel on vehicles that makes it universal; useful
certainly, but not universal. Rather it is the concept of the
wheel rotating around a central axle that is its universal
aspect.

In fact the invention of the potter's wheel may pre-date the
vehicle wheel, and some have proposed a date as early as
8,000 BC. The potter's wheel enables much larger and
higher-quality pots to be manufactured than can be

constructed using clay coiling techniques and just one's hands. The universality of the wheel doesn't stop with the potter's and vehicle wheel though; there's also the spinning wheel, the water wheel, the millstone, the cogwheel, and more recently the flywheel and turbine. In fact, once you start looking at our world you see wheels everywhere and in all sorts of things.

A test for *universality* in an anthropological sense is to consider what would happen to a society if the universal tool or machine was suddenly and completely removed. If Stone Age man woke one morning and all the hand axes had vanished, along with the flints, so no more could be made, he'd find butchering the animal he killed yesterday very hard. The family might have to resort to trying to bite bits of raw meat off the bones. They'd not be able to chop small trees for wood or make new spear points. Making new leather would be very difficult and cutting existing skins would be almost impossible. All sorts of tasks that were trivial the day before, when they had hand axes, would be difficult and take much longer; their productivity would have plummeted. There would be no time for gazing at the stars that night and telling stories around the fire.

If we move on to the wheel, we could consider any society from the Bronze Age to modern times to see what impact the sudden and total loss of the wheel would have. Let's imagine it's the Victorian era, and people woke up and all the wheels had vanished. There would be no trains or horse-drawn carts and carriages; no delivery system really, except for canal boats. There would be no industry, no textiles, no mining, and no machinery of any sort. No clocks, no newspapers to report that the shops were running out of food and no way to even grind corn. Society would come to a complete standstill without the wheel.

So let us now apply this measure of universality to the computer and wake up in a world without computers of any kind. You'd probably over-sleep that day because it would be unusually quiet and your alarm clock would not have worked if it's digital (the computer chip in the clock would have disappeared). There'd be no electricity without computers to control the generation and distribution processes. There may be no water, and if there was it would soon run out, as the water utility pumps are all computer-controlled. There would be no radio or TV to tell you what was wrong, and both landline and mobile phones would not work. Of course, the Internet wouldn't exist. The only communication system would be word of mouth or carrier pigeons.

All the banks would be closed and no credit card or electronic payment systems would be working. Unless you had cash, you would be unable to buy anything, and even then most stores would be closed because most staff wouldn't get in to work. Cars wouldn't start, because all modern cars have engine management and braking systems that use computers. Buses wouldn't be operating for the same reason, and other public transport systems like trains, subways and trams are even more reliant on computers. In fact, a bicycle may be the only way to get around other than walking. If you did find a store that was open, it might even be unwilling to sell you anything, as its stock control and inventory systems would be down and even the cash registers wouldn't be operating.

So you see, it's quite scary to realize just how universal the computer has become; how totally dependent our society is on these machines. Our society truly doesn't function without computers. The purpose of this book is to tell the story of how we got to this point, and imagine what lies ahead for us and our newest and most useful universal machine—the computer.

Chapter 2

THE DAWN OF COMPUTING

We're all quite familiar with what a computer is. You probably have one at home, and you may also have one on your desk at work, or to provide some other work function such as a cash register or for stock control. Basically, your computer at home or work is a box connected, usually by cables, but sometimes wirelessly to a mouse, keyboard, screen, and perhaps a printer and scanner. Every computer you've ever seen has probably looked much like this. However, as we will see in this and subsequent chapters, a computer need not look like this; indeed, a computer need not look like anything at all. It can even be an abstract mathematical concept with no physical reality. The computer we will look at in this chapter doesn't look like your PC or Mac; in fact it's not even electronic—it's totally mechanical.

THE DIFFERENCE ENGINE

In 1985 an Australian computer scientist, Dr. Alan Bromley, had been studying the original plans for a Victorian mechanical calculating machine called the "*Difference Engine No.2.*" He concluded from the plans that the Difference Engine could be built and would work. Bromley contacted Doran Swade, of London's Science Museum, and they decided to build a trial part of the Engine to test if the complete machine could be built. Assuming the trial part was a success, the 200th anniversary of its inventor's birthday on the December 26 1991 would make a fitting date to unveil a fully working Difference Engine.

I. Watson, *The Universal Machine,*
DOI 10.1007/978-3-642-28102-0_2,
© Springer-Verlag Berlin Heidelberg 2012

The production of the trial piece pushed the technical skills of the Science Museum's engineering workshops and machinists to their limits; work proceeded slowly. Although the engineering drawings were complete, many details had been omitted such as the precise shape of teeth on the numerous cog-wheels, the materials the pieces were to be constructed from, and even the function of some elements in the design were baffling. Bromley estimated that a commercial build of the complete Engine by an engineering firm would cost around £250,000. The Science Museum was never going to fund this and neither was Margaret Thatcher's government, so commercial sponsorship was needed.

Doran Swade, the new *Curator of Computing* at the Science Museum, was planning a new computing and telecommunications gallery for the museum. A working Difference Engine, he thought would make a wonderful central exhibit, firmly positioning an Englishman as the founder of computing. *ICL* (a leading UK computer company), *Hewlett Packard*, *Xerox*, *Siemens* and *Unisys* came up with the necessary sponsorship and the build was on.

There were many problems because of the lack of detail in the design drawings, compounded by the obvious fact that the designer was long dead and couldn't be asked questions. For example, were the moving parts to be lubricated? If they where with what, and how? If the machine parts were lubricated, how would the lubricant be changed and cleaned, since oil and grease gathers dust that can eventually clog mechanisms. For this reason many precision instruments such as clocks and watches are sealed and designed to run dry without lubrication. The builders of the new Difference Engine had to solve these and numerous other design issues. They also found some errors in the design that had to be corrected. However, they firmly believed that had its inventor built his Engine, he would have realized his mistakes and easily corrected them. The design was inspired—an elegant thing of beauty.

Forty-six subcontracting engineering firms were contracted to manufacture 4,000 components. The Engine would sit on rails 11 feet long and be constructed in public view in the

Science Museum. The *"cast iron computer,"* as it was nicknamed by the press, started to gain media interest and its inventor's bicentennial was now only months away.

A major problem was encountered as the build neared completion and elements of the giant three-ton machine could be tested. Getting all the gears and cogs in correct alignment to start the machine was a nightmare; if the alignment was even slightly out, the machine jammed. Moreover, some cog teeth, cast in bronze, were snapping off and flying dangerously, like shrapnel, from the machine; hardly a desirable feature for a family-friendly museum exhibit.

The Difference Engine No.2 in the Science Museum, London

On the opening day of the exhibition, the Difference Engine No.2 stood proudly at the center of the display, its mahogany plinth gleaming and its oiled bronze and steel reflecting the bright lights of cameras. It's a truly wonderful creation: the ultimate *steam-punk* fantasy, a marvel of Victorian engineering and of one man's vision. A member of the build team stands ready to crank the handle to start calculating. What the watching lenses of the media don't know as the machine springs beautifully to life, is that it has been, in effect, put out of gear and into neutral. The handle turns and it seems to come to life, but the Engine is actually idling. The build team can't risk a public failure in front of the media, so they fake it for the news cameras.

Almost a year later, by methodically ironing out all the small problems; increasing the precision of some of the components to the same exacting tolerances that obsessed its inventor, and even hand-finishing some components, just as had been done over a century before, the Difference Engine finally works. On Friday November 29 1991, 120 years after its inventor's death, the Difference Engine produced its first totally error-free, fully automatic calculation and then repeated and repeated the calculations, over and over again, without any errors. The Difference Engine works!

THE FIRST COMPUTERS

The first computers were not mechanical; they were people. *Computers* in the nineteenth century were people who performed calculations for mathematical tables as a job, just as *typewriters* were people who typed for a living and *printers* were people who printed. Being a computer was a particularly tedious, boring and difficult job. Even up to the middle of the last century, a computer was defined in dictionaries as a person:

> **Computer**: *"one who computes; a calculator, reckoner; specifically a person employed to make calculations in an observatory, in surveying, etc." Oxford English Dictionary, circa 1945*

These tables were vitally important; celestial and lunar tables were used by navigators to plot a ship's position on the seas. Surveyors, engineers and architects used tables in their calculations when designing buildings and bridges, whilst gunners used ordinance tables to calculate the trajectories of shells from cannon. Mistakes in these tables could therefore have serious, even deadly consequences. An example of a table from a modern celestial almanac is shown below.

A page of the 2002 Nautical Almanac published by the U.S. Naval Observatory

The people who were computers had some education, but were by no means mathletes, nor did they need to be—the calculations were typically mundane but highly repetitive. In England, computers were often retired clergymen or school teachers. In France, after the revolution, many were ex-wig makers. They had watched their wigs—and the aristocratic heads upon which they rested—being lopped of by Monsieur Guillotine's wonderful new execution machine. Now unemployed, as the number of aristocratic heads literally fell, they became computers.

For accuracy's sake, two *computers* were routinely set the task of independently calculating the same table so their results could be cross-checked for errors by a third person, called a *comparator*. Once checked, the tables would be transcribed into a single final version and then sent for typesetting and printing. After printing, the table would then be compared again with the final transcribed table and only if no discrepancies were found, declared correct. Unfortunately, errors could be introduced at each of the four separate stages: computation, transcription, typesetting and proofing. Typesetting was a particular source of errors, as individual numbers (in mirror image) were laid out into printing blocks. When English is typeset, the typesetters, who are adept at reading in mirror image, can easily spot misspelled words or errors such as letter transposition. For example, look at this quotation from the *US Declaration of Independence*:

> *"We hold these truths to be self-evident, that all nem are created equal, that they are endowed by their Creator with certain unalienable Rights, that among these are Life, Liberty and the pursuit of Happiness."*

Spotted the error? Most of us can spot such typos easily. Now look at this sequence of numbers, which is the value of π (Pi) to 50 decimal places:

3.14159 26535 89793 23846 26433 83279 50288 41971 96399 37510

Of course you easily spotted the error and know that the correct sequence should be:

3.14159 26535 89793 23846 26433 82379 50288 41971 96399 37510

Can you even spot the difference? Now you understand the problem: manually checking page after page of numbers for errors is very, very, very difficult—in fact impossible. Thus, all of the mathematical tables in routine use had unidentified errors in them, which could lead to a bridge being incorrectly engineered or a ship's captain believing his ship was in a totally different position to its actual location.

Babbage was working with a friend in laboriously comparing the tables produced by two computers when he reportedly said:

> *"I wish to God these calculations had been executed by steam."*

By which he meant that he wished a machine existed that could produce the tables automatically. Babbage then decided he would build such a machine.

THE GRANDFATHER OF COMPUTING

Charles Babbage in 1860

Babbage was born in 1791 in London, the son of a wealthy merchant banker. He spent most of his youth in the West Country of England, where his father's wealth enabled the young Babbage, who was a sickly child, to obtain a good education and develop a deep love of mathematics. He eventually went to Trinity College and then to Peterhouse College, Cambridge, but he was very disappointed to discover that the education there was not advanced mathematically and that he was better educated in the new continental math than his tutors. To counter the lack of advanced mathematical teaching and research at Cambridge University, he and several life-long friends, including John Herschel (more of whom later), formed *The Analytical Society* in 1812.

Although Babbage was a brilliant mathematician, he failed to graduate with Honors because of either his stubborn independent streak or a perverse wish to destroy his career. Failing to graduate with the highest Honors would ruin the career of a young mathematician, making it impossible to get good university or government positions. The Honors exam at Cambridge was a complex ancient process, called the *Acts*, culminating in an oral, *viva voce*, exam. The candidate had to propose a thesis that would then be attacked in a formal public debate. Babbage proposed a thesis to defend that stated, "*God was a material agent.*" The moderator of the debate, one Reverend Thomas Jepherson, judged the proposition blasphemous and Babbage was promptly failed.

This result was entirely to be expected; the great majority of university academics of the time were theologians and were deeply religious. Babbage must have known beforehand that his thesis would get him shot down in flames and could not succeed. His best friend, John Herschel, advised him so, and went on himself to graduate with Honors and then get a prestigious government job. Babbage's largely self-imposed academic purgatory was the start of a life-long obsession with external recognition, and an iconoclastic war on the powerful political and scientific establishment of England. Babbage graduated in 1814 with just an *ordinary* degree.

Not content with ruining his academic reputation, Babbage then proceeded to secretly marry Georgina Whitmore, the sister of a close friend, against his father's express wishes. Babbage's father did not cut him off however, and continued to provide him an annual allowance of £300, which, though not a fortune, enabled the newlyweds to live comfortably. Still, the young Babbage had no career. He applied for several university positions, but his lack of honors and of political patronage counted against him.

Job appointments in the early nineteenth century were rarely awarded on merit. Most appointments were seen as political gifts to be bestowed upon those who lobbied the hardest or who had the most powerful friends and connections. Babbage railed against the inequality of this system of patronage, perhaps because his father, though wealthy, was not well connected. Georgina's father died in 1816 and with her inheritance their annual income increased to a very comfortable £2,000 a year.

During this time Babbage continued with his mathematical studies and other scientific interests, and began to establish himself within the scientific circles of the time. In 1815 he gave a series of lectures on astronomy at *The Royal Institution* that were well received, and he was elected a *Fellow of the Royal Society*. In 1820 he and some friends, including his college friend Herschel, founded *The Astronomical Society* in a pub in Lincoln's Inn Fields. In 1821 Babbage and Herschel went on holiday together to the Italian Alps, and it was during this trip that the genesis of the idea of a calculating engine was born.

Babbage and Herschel were preparing a new almanac of logarithmic tables and were painstakingly cross-checking the tables produced by their two *computers*. Babbage was familiar with the whole process involved in printing the almanacs and must have known that even if he and Herschel produced perfect logarithmic tables, they could still be corrupted during the typesetting process.

Babbage set about designing a mechanical *engine* that could calculate and, most crucially, automatically print the mathematical tables. He called his invention the *Difference Engine*. Soon Babbage had made a miniature working model of his Difference Engine, and had given much thought as to how the result could be automatically printed by impressing the results into soft metal or papier-mâché that could then be used as molds for printing plates. By having a seamless process from start to finish, with no human intervention, all errors could be completely eliminated from the process.

But how can a calculating engine solve complex equations such as those used to create the results in the almanac tables? The Difference Engine was an ingenious device, being able to calculate sequence after sequence of results for a table using a well-established technique called *finite differences,* which enables complex calculations to be made just by addition without the need for any multiplication or division. The basic technique is really quite simple but feel free to skip ahead to the next section if you're not interested in the details.

Perhaps, surprisingly to you, all actual calculations contain errors because there are limits to how much precision you can have in any real measurement. If you calculate the diameter of a circle using Pi, you will probably only use Pi to an accuracy of two or three decimal places. Even if you use Pi to 50 decimal places, that is still not as accurate as using 100 decimal places or 1,000. In 2002 a Canadian team (using a very powerful computer) calculated Pi to an accuracy of 1,241,100,000,000 decimal places, but even this is still not perfectly accurate, since Pi has an infinite number of decimal places. If an engineer measures the arch of a bridge, she can't use infinite decimal places; therefore, in real world calculations we start with approximate values that are thus inherently imprecise.

Most mathematical formulas can be approximated via polynomial series such as:

$$a_n x^n + a_{n-1} x^{n-1} + \ldots + a_1 x + a_0$$

Let's see the values for the simple polynomial: $2x + 1$.

$$\text{If } x = 1 \text{ then } 2x + 1 = 3$$

$$\text{If } x = 2 \text{ then } 2x + 1 = 5$$

And so on. We can put this into a table where **f(x)** is the result of the polynomial function and **diff** is the difference between successive results.

x	f(x)	diff
1	3	
2	5	2
3	7	2
4	9	2

The first column is the value of x; the second $f(x)$, the result of calculating the result of our polynomial function $(2x + 1)$. The last column **diff** is the difference between the values $f(x)$ of the current row and $f(x)$ of the row above. Notice that in the table above **diff** is always 2. In fact, for any polynomial of degree $n = 1$, the difference is always a constant.

Now let's look at a more difficult polynomial where $n = 2$, such as: $(2x + 3x) \times x$, we could construct a difference table like this:

x	f(X)	diff1	diff2
1	5		
2	20	15	
3	45	25	10
4	80	35	10
5	125	45	10
6	180	55	10

In this case, the first column of differences (**diff1**) changes in value, but if we calculate the difference between their successive differences we get the values for **diff2** that now are all the same (10). In fact, for any polynomial of degree 'n', the 'nth' difference is always a constant number.

Babbage's Difference Engine works by using this principle. Once the engine is given all the values for a row, it can calculate any number of subsequent rows by simple additions. In this last example, you only need to have values of the first complete row ($x = 3$) to be able to easily generate by addition all the rest of the table. Consider $x = 4$: we know its value of *diff2* is the constant 10. We can calculate the value of its *diff1* by adding 10 to the *diff1* value of $x = 3$ to give us 35. We can calculate the value of $f(x)$ for $x = 4$ by adding 35 (*diff1* for

$x = 4$) to $f(x)$ of $x = 3.35 + 45$ equals 80, thus $f(x) = 80$. Now you can calculate the values for $x = 7$.[1]

x	f(X)	diff1	diff2
1	5		
2	20	15	
3	45	25	10
4	80	35	10
5	125	45	10
6	180	55	10
7	?	?	10

Okay, so we can calculate difficult polynomial equations by just using the method of difference and simple addition; but what about other mathematical equations? Here's the useful trick. Almost all regular mathematical expressions such as logarithmic and trigonometric functions can be approximated by a polynomial to a given degree of accuracy. Remember our discussion earlier, that all our measurements are approximations. So you see, we don't need to calculate our engineering or astronomical equations precisely. It's okay to approximate if we know the degree of accuracy. Moreover, if we can be certain that the approximation is completely error-free, then this is a huge advance.

Babbage knew that if a Difference Engine was provided with a complete initial row of numbers, it could calculate the value of any polynomial function that approximated other more useful functions to 20 decimal points of accuracy just by using simple additions. He imagined a machine (or *engine* as he preferred to call it) that at the simple turn of a crank handle

---

[1] For $x = 7$, $f(x) = 245$, $diff1 = 65$ and $diff2 = 10$

would calculate the next value for *f(x)* without the possibility of error. If the crank was attached to an engine such as a steam engine, and if the results could be embossed onto an engraving plate, then the entire procedure could be automated. Babbage believed that such an Engine would be of great scientific and economic advantage to Britain, since it would be the only country with truly 100% reliable scientific tables.

BABBAGE GOES PUBLIC

In June 1822 Babbage wrote to the Astronomical Society and described his machine, claiming it produced results "*almost as rapidly as an assistant can write them down*". At this time the French, England's great rival, were producing a definitive set of logarithmic and trigonometric tables using the new Napoleonic metric system. The set of tables would run to 18 volumes and was compiled manually by a semi-industrial process involving dozens of computers and comparators, using the difference method described above. Babbage wrote an open letter to the President of *The Royal Society*, Sir Humphrey Davy, claiming that his Engine could automate the task and moreover would produce totally error free tables, unlike the French. Babbage clearly believed it to be of national importance for England to calculate faster and more accurately than the French. This would not be the last time that computers would be seen to be important to national security, as we shall see in Chapter 4.

Babbage had copies of this letter printed and sent to influential people, and eventually Sir Robert Peel, First Lord of the Treasury, took interest (yes, the same man who invented the police force). It was decided that The Royal Society should investigate Babbage's engine. A committee was formed, of which many members were Babbage's friends, and they reported in May 1823, commending the machine and concluding that Babbage, "*was highly deserving of public encouragement in the prosecution of his arduous undertaking*". However, Babbage was not without powerful enemies, including the Astronomer Royal, George Airy. He later wrote privately to the Chancellor of the Exchequer stating that the members of the committee, "*were all private friends and admirers of Mr. Babbage*" and, "*I cannot help thinking that they were blinded by the ingenuity of their friend's invention*".

However, the Astronomical Society (remember, Babbage was one of its founders) awarded him its Gold Medal in recognition of his invention, and the government announced it would provide financial support to construct an Engine. An initial Treasury payment of £1,500 was issued. Armed with funding, Babbage set about designing a full-scale machine and toured the engineering centers of the country to learn about the latest mechanical techniques, materials and processes.

Babbage's father died in 1827, leaving Babbage a fortune estimated at £100,000. He was now free of any responsibility to earn a living, and could live as a gentleman philosopher and enjoy high society life in London. He and his wife had eight children, of whom only three lived to adulthood. Babbage lived the life of a gentleman of independent means in prosperous Portland Place, London, and dabbled in all aspects of science, eventually being elected to *The Royal Society*. By modern standards Babbage's scientific activities would seem prodigious, since he had an active interest in mathematics, engineering, cryptography, geology, and many other areas. However, this was not uncommon in Victorian England when, unlike today, an educated person was expected to be an expert in many different scientific disciplines.

His interest in cryptography and ciphers led him in 1854 to crack the *Vigenère Cipher*, which had remained unbroken for over 300 years. However, Babbage's triumph was not publicized for nine years, until Frederich Kasiski cracked it independently. It is believed Babbage's breakthrough was kept secret for military reasons, as the British were at war with the Russians in the Crimea and the Russians were using the Vigenère Cipher to encode their messages, which the British could now read.

Babbage also invented the *cowcatcher* for the front of railway locomotives to deflect obstacles from the front of trains and reduce the risk of derailment. Another invention of his for the railways was a *black-box* that could record the motion of a train should there be a crash. So although the development of the Difference Engine was not Babbage's sole interest, it was however developing nicely. The main problem he faced was not of design, but rather of engineering and construction.

The basic problem was that, unlike today, there were no engineering standards. Babbage could not, for instance, order 100 brass bolts of 15/32" gauge with matching hexagonal nuts from a catalogue. All such things were hand made by individual craftsmen and foundries as bespoke items. A bolt purchased from one manufacturer would not work with a nut purchased from another. This meant that every piece of the Difference Engine would have to be custom-made. Babbage's friend, the famous engineer Isambard Kingdom Brunel, recommended a talented draftsman and machinist, Joseph Clement, who had gained a reputation for excellent precision work, and so Babbage hired him.

Precision was key in the design and build of the Difference Engine. Babbage would tolerate no errors in the calculation, since an error early in the preparation of a difference table would then propagate through the entire calculation. Complex error-checking mechanisms were built into the design that was pushing the limits of Victorian engineering to new heights. The Engine could calculate to 50 digits of precision, but its design comprised hundreds of identical wheels and cogs.

Today a computer-controlled lathe would automatically manufacture such pieces to within a micron of accuracy. In Babbage's time, these pieces had to be individually cast or milled and then finished by hand to ensure, for instance, that two inter-meshing cogwheels had a precise accurate fit. This was a laborious and expensive process. The Engine had approximately 25,000 individual parts, was eight feet high, seven feet long and three feet deep! Fifteen tons of handcrafted high precision components; a mechanical build of this complexity had never been attempted before.

1827 was a momentous year in Babbage's life. The death of his father affected him deeply, even though their relationship had been a stormy one. Later in the year his young son, also called Charles, died and then in October his wife, Georgina, died in childbirth along with his newborn son. Babbage was devastated, and near physical and mental breakdown. On the advice of friends, he left England for a continental tour.

He obtained leave from the government to suspend work on the Engine, and spent over a year touring Europe. Although he had received £1,500 from the Treasury the design and build, he estimated, had already cost him over £3,000 (and this would be discounting the estimated £3,000 he had spent developing the initial design). With hindsight, perhaps, requesting complete reimbursement of expenses for the project whilst on an extended European holiday, with the project on hold, was not the most politic idea.

Five or six years after the project had started, and with nothing to show for their money, people started questioning whether further huge sums should be committed. Letters were written to the *Times* newspaper, and Babbage's detractors even insinuated that monies had been misappropriated.

On a positive note, Babbage was elected to the *Lucasian Chair of Mathematics* at Cambridge University. This is the position that Sir Isaac Newton had held, and which Professor Steven Hawking held until recently. It seems strange to us now that Babbage never felt he achieved the recognition and distinctions that he deserved, when he would be in the same company as Newton. In fact, Babbage actually considered declining the appointment, which only paid £100 per annum, but appeared to have no required duties. He didn't even have to live in Cambridge or give any lectures. However, on the advice of friends, he accepted and held the post for a decade. Afterwards he commented bitterly that it was, "*the only honor I ever received in my own country.*"

Babbage was, understandably, deeply upset by allegations of financial impropriety and wrote to the Prime Minister, the Duke of Wellington, who passed the buck to the Royal Society, which, you guessed it, appointed a committee, chaired by Babbage's old friend Herschel, to investigate. Not surprisingly the committee's report totally vindicated Babbage and a further £1,500 was obtained from the Treasury.

The financial insecurity of the project had caused problems with Babbage's relations with Clement; by all accounts a difficult man to deal with at the best of times. The relationship between the two men had developed poorly and it had become difficult to tell what was Babbage's original design and what was Clement's development. This was particularly true for the machine tools that Clement had developed to fabricate the components.

Babbage now tried to formalize their relationship and claimed all intellectual property in the drawings and the tools, whereas Clement, as was customary at the time, claimed ownership of his tools. He also claimed that he was still owed £2,000 despite having already been paid £3,000. Clement ceased work in May 1829 and the pair were forced into arbitration. Clement was paid in full and work resumed, but their relationship had been badly damaged.

Babbage now estimated that it would take three years to complete the build, and once again he applied to the Treasury for more funds. Once again it was referred to the Royal Society, which endorsed the proposal. Because of the size of the Difference Engine, a new building was constructed where Clement and his family could live on-site. Clement was also instructed to build a small section of the Engine so that Babbage could demonstrate its function, and perhaps reassure the Treasury that it would work. In 1832 the demonstration section, about two feet square, was installed in Babbage's house.

Clement had never been happy with the move to new premises, since it meant he had to close down his own workshop, and he had claimed £600 in compensation. After a prolonged and bitter dispute, Clement's claim was rejected and in March 1833 he downed tools again, fired all his employees and walked off the project. A complex wrangle then ensued over the drawings of the Engine, the machine tools and components, involving Babbage, Clement and the Treasury. So far the project had cost almost £17,000—a huge sum of money in the day. By comparison, less money had been offered as the prize for the solution to estimating longitude accurately on ships.

It took a whole year for the dispute with Clement to be resolved and all drawings, tools and parts to be returned to Babbage. During this time Babbage was demonstrating the small working section of the Engine at social functions in his house, to the wonder of guests. This *"finished portion of the unfinished engine"* was the only piece to be completed in Babbage's lifetime.

THE ANALYTICAL ENGINE

During the long delays in construction of the Difference Engine, Babbage had started to revisit his design, and perhaps playing with the working section had made him rethink his original ideas. The notion of calculating based on differences was elegant and efficient, but limiting. Babbage was starting to envisage a general-purpose calculating engine that could perform any calculation—a more *universal* engine. In particular he was starting to think about how the product from one calculation could be fed as an argument into another subsequent calculation. In this way complex computations could be built up from simpler components.

He tinkered with the working section of the Difference Engine to allow a digit from one column to be fed automatically to the input wheels of another. With a leap of insight, he redesigned the linear array of calculating columns in the Difference Engine into a circular layout where the last column could feed its results back to the first. Babbage referred to this as, *"the Engine eating its own tail."*

Babbage wrote: *"The whole of arithmetic now appeared within the grasp of mechanism. A vague glimpse even of an Analytical Engine at length opened out, and I pursued with enthusiasm the shadowy vision."* From 1834 to 1836 Babbage invented automated mechanisms for multiplication and an ingenious, though complicated method for mechanical long division. Babbage became almost totally fixated on the design of the Analytical Engine and employed new staff to help him produce the technical drawings. Work on the Difference Engine had now completely stopped.

Babbage was highly knowledgeable in aspects of advanced machinery, and in particular was inspired by complex textile manufacturing equipment; state-of-the-art machines of their time. Some of the terms used to describe the Analytical

Engine are derived from textile manufacturing. As a computer science professor, I have often told students that the Analytical Engine has the same architecture as a modern computer.

The Analytical Engine is a computer made of component parts. *The Store* contains numbers to be processed or operated on—we would now call this memory. *The Mill* processes the numbers from the store—we would call this the central processing unit (CPU). In the Analytical Engine there is both a logical and physical separation between memory and CPU. Babbage also envisaged input and output devices that were remarkable prescient: punch card readers to provide calculating instructions to the Mill (i.e., programs to the CPU) and numbers to the Store (data to the memory) and output to printers using carbon paper, or to punch cards. The use of punch cards was directly borrowed from weaving machines, like the Jacquard Loom that used these cards to input the complex patterns for lacework.

Punch cards for a Jacquard Loom

Although punch cards are no longer used by computers they are still used by some modern knitting machines.

So during a couple of years in the mid-1830s, Babbage had designed a machine that could automatically add, subtract, multiply and divide numbers 50 digits long. These simple arithmetic operations could be combined into complex calculations that could include iterations or loops as well as conditional branching (if ... then statements). The machine could read its instructions in from punched cards and output its results to punch cards that could be stored and used later, or it could directly print its results in duplicate. The Difference Engine could perform one calculation over and over reliably, but the Analytical Engine could in theory perform any calculation that could be programmed into it. The Difference Engine was a calculator; the Analytical Engine was a computer.

The Analytical Engine would have been a colossus! The Mill was 15 feet tall and six feet in diameter, whilst the size of the Store depended on its capacity (just like modern memory). To store 150 numbers would require about 20 feet of Store. Babbage envisaged a Store for a 1,000 numbers over 100 feet long. The Analytical Engine was never built, but it would have been one of the engineering marvels of its time; right up there with Isambard Kingdom Brunel's Great Western Railway, the Clifton suspension bridge and the Great Eastern steamship.

Trial model of a part of the Analytical Engine, built by Babbage

Somewhat understandably, Babbage had now lost interest in the Difference Engine, but the stalled project was a constant reminder and given the huge sums of public money sunk into it, he was worried as to what his duty to finish it might be. In December 1834 he wrote to the Duke of Wellington, now Foreign Secretary, stating that he had developed *"a totally new engine possessing much more extensive powers"*. He latter clarifies this letter and states that, *"it would be more economical to construct an engine on the new principles than to finish the one already partly executed"*. Effectively he is saying that the Difference Engine was obsolete. The government acted rather calmly to this revelation and there were no immediate repercussions. This of course may have been a simple oversight, since there were four chaotic changes of government within months of each other at this time.

You may think that since Babbage had little interest in completing the Difference Engine, which he now thought obsolete, that he'd be happy to let sleeping dogs lie, and that if the Government in its confusion had taken its eye off the ball, then he'd be well advised to let the aborted project quietly drift into history. But no, in 1842 Babbage started writing to the new Prime Minister, Robert Peel. The Prime Minister had some rather pressing issues on his agenda, like rioting and starvation in Ireland, and an industrial depression, and he eventually passed on the issue of what to do about the Difference Engine project to the Chancellor. Rather than consult with the Royal Society and Babbage's old friend Herschel as before, the Chancellor asked the Astronomer Royal, George Airy, to advise. Airy had never seen the utility of the Difference Engine. In particular he did not see the need to automatically generate navigational tables when they had been economically produced for years by hand. Airy wrote:

"The necessity for such new tables does not occur, as I really believe, once in fifty years. I can therefore state without the least hesitation that I believe the machine [the Difference Engine] to be useless, and that the sooner it is abandoned the better it will be for all parties."

"On Sept. 15th Mr. Gouldburn, Chancellor of the Exchequer, asked my opinion on the utility of Babbage's calculating machine, and the propriety of expending further sums of money on it. I replied, entering fully into the matter, and giving my opinion that it was worthless."

This was damning stuff, since it went to the heart of the need for *"the machine,"* not the likelihood of successfully building it. Airy was arguing that it didn't matter if the Difference Engine worked or not; it just wasn't needed.

Gouldburn wasted no time, and wrote to Babbage saying that the Government would spend no more money on the Engine, and that it would have no claim on it whatsoever. However, Babbage was free to do what ever he wished with the partially completed Engine. Now we can understand that Babbage would be upset by the official axing, after 20 years, of his life's work. We'd also expect him to be angered by Airy's dismissal of the Engine. However, it seems that Babbage completely lost the plot, even though he didn't really want to continue the development of the Difference Engine anyway.

He demanded a meeting with the Prime Minister, although he'd been advised by a friend that he should not anger Peel because he was exhausted with the demands of office during very turbulent times. Babbage went into the meeting with all guns blazing, demanding reimbursement for all the time he'd lavished on the project to date, and wanting payment in advance for the completion of the Engine and the right to own the copyright on all tables produced by it in the future. He also demanded a civil honor (such as a knighthood) and perhaps a state pension to prove that he had acted honorably throughout the project.

Peel, not in a good mood anyway, was apparently very angry. Even Babbage says so, and he was kicked out of Downing Street. The Difference Engine project was now quite dead.

THE ENCHANTRESS OF NUMBERS

Around this time, a woman enters our story. Ada, the Countess of Lovelace, sounds like a character from a bodice-ripping novella, but she was actually the daughter of Lord George Byron, the celebrated romantic poet. She was for her time rather an unusual young woman and perhaps, as a consequence, a lot of nonsense has been written about her. So let's first kill a couple of the untruths.

Ada was not the world's first computer programmer, even though the programming language Ada is named after her. She was not Babbage's patron; he was after all a wealthy man in his own right and the government had been funding the development of the Difference Engine anyway. However, she was a remarkable woman for her time, with a keen interest in science and mathematics in particular.

Remember that in times past, women's minds were held inferior to men's and it was widely thought they were only good for looking after a home, raising children and perhaps embroidery and pretty piano playing. As Jane Austen observed, "*A woman, especially, if she have the misfortune of knowing anything, should conceal it as well as she can.*" However, Ada believed, "*The more I study, the more remarkable do I feel my genius for it to be.*" Ada's mother, Lady Byron, had been mathematically educated. Byron referred to her as his "*Princess of Parallelograms*" and encouraged his daughter to be similarly educated.

Ada the Countess of Lovelace

Ada moved in the same elevated social circle as Babbage, being married to the Earl of Lovelace, and she became fascinated with the Difference Engine and the potential of the Analytical Engine after meeting him at a party and later seeing a demonstration of his "*Thinking Machine.*" She understood what a breakthrough to science and engineering a reliable computer would be.

In 1840, Babbage had visited Italy and given some lectures in Turin that had been very well received. An Italian engineer, Luigi Menabrea, had subsequently published a description of the Analytical Engine in Italian. Ada translated this paper into English, and at Babbage's request added comments of her own. In collaboration with Babbage she worked vigorously on the project, signing herself in correspondence as "*Babbage's*

Fairy." However, once again Babbage was to fall out with a collaborator.

Ada was keen to gloss over the troubles that Babbage had with the building of the Engine and with the government, whereas Babbage saw publication as an opportunity to vindicate his position and his reputation. In the end their writings were published separately; hers in *Scientific Memoirs* and his in *The Philosophical Magazine*. Ada*'s Sketch of the Analytical Engine* was very well received, and her lengthy additional notes to Menabrea's original description gave her the freedom to muse on the Engine's more philosophical aspects. She comments that:

> *"The Analytical Engine weaves algebraic patterns just as the Jacquard loom weaves flowers and leaves... Supposing for instance, that the fundamental relations of pitched sounds in the science of harmony and of musical composition were susceptible of such expression and adaptation, the Engine might compose elaborate and scientific pieces of music of any degree of complexity or extent."*

Ada was the first person to hypothesize that a machine, by manipulating symbols, could perform a task like composing music. In essence a machine, like the Analytical Engine, could perform any task that could be expressed in symbols. It would be a universal machine.

> *"Many persons...imagine that because the business of the Engine is to give its results in numerical notation the nature of its processes must consequentially be arithmetical and numerical, rather than algebraical and analytical. This is an error. The engine can arrange and combine its numerical quantities exactly as if they were letters or any other general symbols; and in fact it might bring out its results in algebraic notation, were provisions made accordingly."*

In her extensive notes to her paper she describes an algorithm for the Analytical Engine to compute Bernoulli numbers. It is considered the first algorithm written for implementation on a computer, but it is far from being a "*program*," more like programmer's pseudo code. It is this that has led people to claim her as the "*first computer programmer.*"

Just as now, they thought about using the machine to play games. Babbage wrote: "*every game is susceptible of being played by an automaton.*" He even invented a machine for playing tic-tac-toe and toyed briefly with the idea of starting an arcade games business. Nobody would realize that a computer was a symbol manipulating machine, and could solve any task that could be represented by symbols, until Alan Turing almost a century later had the same remarkable insight.

As an aside, it is also worth noting that it is certain that Ada had read Mary Shelly's Gothic story *Frankenstein*, which was extremely popular. Mary Shelly was the poet Percy Shelly's wife, and Shelly was the best friend of Ada's father, the "*mad and bad*" Lord Bryon. In Frankenstein, medical science and electricity is used to create a conscious being, the monster, who eventually turns on his creator because he can't find love. Ada and Babbage were thinking about using mathematics and engineering to create a machine intelligence. The *Frankenstein myth* is at the heart of much science-fiction, most notably in *2001 A Space Odyssey*, *Blade Runner* and the *Terminator* series. The Frankenstein myth and computers were thus intertwined at birth, and we will probably always wonder if our creations may turn on us in the future.

Tragically, in 1852, Ada died of cancer at the age of just 36 and is buried next to her famous poet father. Although she was a very remarkable woman for her time, writing the most well-known and influential article describing the Analytical Engine, historians are wrong to believe that she helped Babbage in any way with its design, and they stretch the truth to say she wrote computer programs for the Engine.[2]

---

[2] Since 1998, the British Computer Society has awarded a medal in her name and Ada Lovelace Day is an organization whose goal is to "*raise the profile of women in science, technology, engineering and maths*" http://findingada.com/

THE DIFFERENCE ENGINE NO.2

Babbage never built the Analytical Engine, but it seems that the design process had inspired him, and between 1846 and 1849 he started work on a new design for a *Difference Engine No.2*. This used the same mathematical principle as the original Difference Engine, but was an improved and much simpler design, using just one-third the number of components to perform the same calculations. Even so, it was still a massive construction; 11 feet long and seven feet high, with 4,000 parts.

We don't know why Babbage returned to the Difference Engine. Was it because he now saw he could improve its design, or was it because he still wanted to satisfy his obligation to the government, which had funded its development? In 1852 he offered the new plans to the government, which were promptly rebuffed and once again Babbage went off in a huff. Perhaps because No.2 was never built and the plans were therefore not used, he left a complete set of 24 detailed engineering drawings that eventually found a safe home in the Science Museum in London.

Babbage's relationship with the establishment suffered another blow. The Great Exhibition of 1851, held in the specially designed Crystal Palace, was the largest exhibition of science, engineering and manufacturing technology ever staged. It was intended to show the world the prowess, might and genius of the Victorian industrial revolution. Six million people visited during its five months. As an eminent Fellow of the Royal Society and a celebrated mathematician, inventor and engineer, Babbage had been expecting to be heavily involved with its organization, perhaps as a Commissioner for Prince Albert. But Babbage was snubbed, probably because he had now earned himself a reputation for being impossible to work with. In response, Babbage dipped his pen in poisoned ink and wrote *The Exposition of 1851: Views of the Industry, the Science and the Government of England*. This bitter piece of work, and in particular a lengthy attack on his arch nemesis the Astronomer Royal George Airy, did Babbage no favors in high places.

A SWEDISH DOPPELGÄNGER

Meanwhile, in parallel to our story a Swedish publisher, journalist and printer, George Scheutz, had been devouring all he could find in print about Babbage's Engines. Despite the fact that the published articles didn't describe the design in detail, George and his son, Edward, set about designing and building their own Difference Engine. Theirs was much simpler than Babbage's; it lacked the complex error prevention mechanisms and was even partly made out of wood. By 1843 the Scheutzes had a working prototype, and in 1851 they persuaded the King of Sweden to invest the modest sum of £270 for further development. The contrast between the Swedish and English experience could not be more pronounced. By the end of 1853, an engine that could calculate four orders of difference to 17 decimal places and automatically print tabulated results was completed, and in 1854 they brought it to the Royal Society in England to display and possibly sell.

Not surprisingly, the Scheutzes were worried that Babbage might be angry that they had copied his ideas; but quite to the contrary he was fascinated and flattered, and welcomed them graciously. It seems that although Babbage was always concerned about getting the recognition he deserved for his inventions, he was a committed member of what we'd now call the *open source* movement—he firmly believed that scientific discovery should be used for the benefit of all mankind and not for personal financial gain. Of course that's an easy line to take if you're already hugely wealthy. Babbage wrote:

> *"They may lock up in their own bosoms the mysteries they have penetrated.... whilst they reap in pecuniary profit the legitimate reward of their exertions. It is open to them, on the other hand, to disclose the secret they have torn from nature, and by allowing mankind to participate with them to claim at once that splendid reputation which is rarely refused to the inventors of valuable discoveries in the arts of life. The two courses are rarely compatible."*

This comment was instigated by his discovery that a friend, William Hyde Wollaston, had kept secret a process to produce malleable platinum for 24 years thereby earning a fortune. As we'll see later, Babbage would have a lot in common with many modern computer scientists.

The Astronomer Royal, though, was not so welcoming to the Scheutz's Engine; once again he put forward his view that despite the accuracy of the machine and the fact that Scheutz's Engine could be produced inexpensively, nonetheless there was no need for it, "*as I believe, the demand for such machines has arisen on the side, not of computers, but of mechanists*". Airy believed that since usable mathematical and astronomical tables already existed and rarely needed to be recalculated, there was no economic need for these machines.

The Scheutz Engine went on from London to the Great Exposition of Paris in 1855, and then returned to London where it was used to produce a 50 page promotional book: *Specimen Tables Calculated and Stereomoulded by the Swedish Calculating Machine*. This was to be the only time the Engine was to be used. Eventually it was sold for £1,000 to the Dudley Observatory in Albany, New York, where it was never used, and now it rests in the Smithsonian in Washington DC.

However, the tale of the Swedish copy of the Difference Engine is not quite finished. During the time the Engine was being shown at the Royal Society, William Farr, the Chief Statistician at *The General Register Office*, became interested in the Engine. Responsible for calculating tables of life expectancy, life insurance and premium tables from census data, he saw the need for automation, particularly since unlike astronomical tables, these had to be regularly recalculated after every census. In 1857, Scheutz and Farr petitioned the Chancellor to provide money to fund a second Scheutz Engine that could be used by three government agencies: *The Royal Observatory, The Nautical Almanac Office,* and *The General Register Office*. Once again Airy was asked for his opinion and he concluded that the Engine, whilst of no use to him or the Nautical Almanac Office, would be of benefit to the General Register. Airy's change of mind rather disproves the view that he just didn't like Babbage and his machines. Once he saw the economic need for the regular and reliable production of actuarial tables, he became a convert to mechanical computing.

The funds were duly granted and a second Engine was built by Donkin & Co. in 1859. The Engine was then used to produce the 1864 edition of the *English Life Table*. However, since Scheutz's design had none of the complex error prevention mechanisms of Babbage's Difference Engine, the machine had to be constantly tended and watched, and its results constantly checked. Babbage's obsession with perfection had proved to be a necessity, not a luxury—an automatic computer was only of use if it was guaranteed 100% error-free. In the end, of the 600 pages of tables, just 28 were produced solely by the Engine and a further 216 partially. George Scheutz died a bankrupt in 1873, and the Scheutz-Donkin Engine can now be seen in the Science Museum in London alongside Babbage's Difference Engine.

BABBAGE'S LEGACY

Ten years after he'd stopped work on the Analytical Engine, Babbage returned to it, dusting off the design and improving sections of it. By 1859, Babbage estimated he could complete the project in about two years, and began ordering components. Four years later it was still unfinished, and by the late 1860s he was still tinkering with the design.

Babbage died on Wednesday October 18 1871 and is buried in Kensal Green cemetery, London. He had published 86 scientific papers, numerous miscellaneous articles and six books, risen to hold the Lucasian Professorship of Mathematics at the University of Cambridge, and dreamed up numerous inventions in areas as diverse as geology, ophthalmics, railways, lighthouses, submarines, electricity, cryptography and even theatre lighting. Even by the industrious standards of his day he had achieved a great deal, yet his life's great work remained unfinished. Neither the Difference Engine nor the Analytical Engine were built during his lifetime.

If you are ever in London, take the time to visit the Science Museum in the South Kensington museum district. In addition to lots of other interesting displays you can see the working replica of the Difference Engine No.2, along with the working part of the original Difference Engine that Babbage had built to demonstrate his ideas. There is also a model of the Analytical Engine, some of Babbage's notebooks and rather oddly half of his brain, pickled in a jar! I visited recently whilst researching this book and I'm now left wondering why they have half of Babbage's brain on display and where's the other half?

After Babbage's death and the failure of the Scheutz Engine, the quest for automated calculating machines died (at least in England) for almost 100 years. Many people speculated that The Difference and Analytical Engines were never built because they were technically impossible. Babbage had been an amateur dreamer, not a serious engineer; a gentleman dabbler, not a professional career scientist.

So what then was Babbage's legacy and what influence did he have on the development of the modern computer? After all he is often referred to as the "*Grandfather of Computing.*" Well, there is some disagreement on this. Many writers cite the architecture of the Analytical Engine, with its separate store for numbers and its processing mill that can be programmed via punch cards, as the direct ancestor of the modern computer. Whereas, in reality the developers of the first electronic computers did not claim to have been influenced by Babbage at all. Of these, only Howard Aitken, who developed the electro-mechanical Harvard Mark 1 at IBM in 1943, claims any influence from Babbage.

Allan Bromley, a historian and computer scientist, states, "*Babbage had effectively no influence on the design of the modern digital computer*". This echoes the verdict of Maurice Wilkes, a distinguished British pioneer of computing, who wrote in 1971 to mark the centenary of Babbage's death, "*[Babbage] however brilliant and original, was without influence on the modern development of computing*". Indeed, Wilkes goes further, arguing that Babbage's costly and very public failure caused the British to shun anything to do with mechanical calculation for almost a century. Perhaps it was this, as we shall see, that caused the Americans to gain a lead in mechanized numerical and information processing. This view was supported by a colleague of Wilkes', L.J. Comrie, who was superintendent of the Nautical Almanac from 1930 to 1936. He used an American Burroughs machine for generating and detecting errors in navigational tables. Wilkes reports Comrie claiming: *"this dark age in computing machinery, that lasted one hundred years, was due to the colossal failure of Charles Babbage."*

The fact that Babbage's Analytical Engine and the modern digital computer share the same architecture should perhaps then be put down as an example of *parallel evolution*, a phenomena well understood in the biological sciences. For example, it is no accident that sharks and dolphins look very similar. They share the same environment and are under similar evolutionary pressures. However, they evolved separately into the similar forms we see today. Thus, the developers of the modern computer had to solve similar problems to Babbage's, albeit with electronics rather than mechanics. As a consequence, they end up with a similar architecture. It is now only with hindsight that we can look to Babbage and think, *"he thought of all this first;"* input and output devices, a programmable central processing unit and separate memory storage for data, and most importantly, a machine that could be programmed to perform any task. A universal machine![3]

_____

[3] As this book was being completed *"Plan28"* was announced to build a working Analytical Engine just as has been done for the Difference Engine No.2. This project is a little harder though as Babbage didn't complete the design of the Analytical Engine and complete engineering drawings do not exist. The team members must first complete the design to Babbage's intentions and then build a working *"steam powered-PC."* More information here: http://plan28.org/

Chapter 3

MARVELOUS MACHINES

Babbage had shown that it was possible to create a machine that could solve any problem—well in theory at least, so of course after his death people set about building versions of his Analytical Engine, right?

Wrong. Actually nobody believed the Analytical Engine would work. It was only an incomplete plan after all, and the Difference Engine had never even been completed, so most people doubted that it would work either. In fact the whole enterprise was seen as an expensive fiasco and probably the obsession of a madman. Nobody was going to attempt to build such complex mechanical calculators again; certainly not in England anyway. The idea of a universal machine was very dead. However, the notion of *specific* tools for *specific* tasks was very much alive, particularly in the United States, where they needed to count their rapidly growing population for the census.

WE THE PEOPLE...

The first US census in 1790 estimated the US population to be almost four million; its population grew rapidly to 31 million by 1860. *The Bureau of Census* was the largest American data processing organization, and for the 1870 census it employed over 400 clerks whose job of tallying up the census results into reports was, like the human computers of navigational tables, unbelievably tedious and slow. When the 1880 census took place the US population was now so large that it took seven years to compile the results. If this trend continued, the 1900 census would take place before the results from

I. Watson, *The Universal Machine*,
DOI 10.1007/978-3-642-28102-0_3,
© Springer-Verlag Berlin Heidelberg 2012

the 1890 census were even published! This problem needed a solution, so the Bureau of Census held a competition to find a mechanized alternative to the tally sheets and manual processes used in previous censuses.

Three inventors entered the competition, and in the fall of 1889 each had to demonstrate their machine's prowess on 10,000 returns from the 1880 census of St. Louis. The competition was won by Herman Hollerith, a graduate of Columbia University, whose machine was much faster than those of his competitors. His *Electric Tabulating System* was adopted for the 1890 system. On June 1 1890 45,000 census enumerators collected the completed returns and sent them to Washington. Six weeks later Hollerith's machines reported the US population was 62,622,250. The Bureau of Census rented 100 Electric Tabulating Systems from Hollerith, who started a business—*The Tabulating Machine Company*.

A Hollerith electric tabulating system

Herman Hollerith's breakthrough came from observing how railway tickets were protected against theft or sharing between several passengers. Each ticket had a *punch photograph* on it where the conductor punched out holes to describe the ticket holder, such as, *female, light hair, short, brown eyes*, etc. In this way a dark haired, tall man couldn't use the same ticket. Hollerith realized that if each question on the census return was recorded as holes in a punch card they could be automatically sorted and counted by a machine.

His machine consisted of two parts: a tabulating machine that could count the holes in a batch of cards, and a sorting box. Each card had 288 holes, and thus a card could contain a maximum of 288 items of data. A press in the tabulator had 288 spring-loaded pins. If a pin met a hole it would pass through, and if not it would be pushed back into the press. The pins passing through a hole dipped into a mercury bath and completed an electric circuit, causing a counter to be incremented. There were 40 counters on the tabulator. The circuit could also cause the sorting box to open for the card to be placed with other similar cards. This operation was done by hand. Operators could process about 1,000 cards an hour, or one every three to four seconds.

The 1890 census was processed much quicker than previously; two and half years instead of seven years for the previous census. The final report was over 26,000 pages long and the whole process cost 11½ million dollars. However, it was estimated that it would have taken a decade to complete manually and cost twice as much.

SOLDIERS AND SECRETARIES

Americans loved office machinery. In fact they liked machines and gadgets of every description, and were proud that they embraced modernity. Unlike their European counterparts they did not have centuries of tradition keeping them stuck to archaic working practices. For example, the *British Prudential Insurance Company* didn't use any office machinery at all until 1915, whereas the *American Prudential Company* embraced every type of automation and eventually in the 1950s become one of the first companies to buy a computer. Many of these early office machines have directly influenced the modern computer, and they automated or speeded up the principal work that business computers still do today: document creation, information processing and calculation.

Wars tend to be very good times for arms manufacturers and the American Civil War had been great business for

Remington. In 1865 the war ended and the US was awash with guns. This wasn't good news for Remington, as nobody was going to need to buy a new rifle for a long time. So management started to look around for a new product that they could manufacture. At its most basic a rifle is a collection of precision-engineered metal components. Thus, Remington's expertise was in making precision-engineered components from various metals and then assembling, distributing and selling them. They needed something with those properties that they could manufacture. In 1873 they found it—typewriters. Remington bought the *Sholes and Glidden* typewriter and renamed it the *Remington No. 1*.

This machine was very much a prototype; it could only type upper-case letters and the typist couldn't even see what was being typed; they were typing blind. Obviously there was room for a lot of improvement. However, despite all the changes to the typewriter there was one feature that remains with us on our computers today—the QWERTY keyboard.

A Sholes and Glidden typewriter, 1976

There's a lot of debate about where the arrangement of the QWERTY keyboard came from. The most popular argument is that if the letters on a mechanical typewriter's keys are arranged in alphabetical order, letters tended to jam when commonly occurring pairs are next to each other on the keyboard. The arrangement of the keys in the QWERTYUIOP manner reduces the number of key jams.

However, this cannot be true because "e" and "r" is the fourth most common letter pairing in the English language[1] Moreover, "r" followed by "e" is the sixth most common pairing, so if you were designing a keyboard to minimize commonly occurring pairs of letters being adjacent, you'd never have "e" and "r" next to each other.

A less commonly held view is that QWERTY was a clever way of promoting brand loyalty. Let me explain, Remington didn't just sell typewriters to businesses; they also ran very lucrative training courses for typists. If a woman (and typists were nearly all women) had been trained on a Remington QWERTY keyboard they would not be able to use any of the competitors' typewriters without retraining, since they had different keyboard arrangements. Moreover, if they went to work for a new business that hadn't yet bought its typewriter, they would naturally insist on a Remington typewriter. Remington quickly established a nationwide chain of training schools and typewriter service centers, and established two key elements of the computer industry to come. The QWERTY keyboard layout we are all still familiar with and also the idea that hardware manufacturers don't just sell equipment. Instead, they sell a complete service, including the machinery, training, servicing and consumables. IBM and other computer companies would later use this business model.

Incidentally, did you know that you can type the word "*typewriter*" using just the top row of a QWERTY keyboard?

In 1886 Remington sold its typewriter business and the new company subsequently merged with *Rand* to become the

---

[1] http://en.wikipedia.org/wiki/Bigram_frequency

Remington Rand Corporation, a major manufacturer of office equipment and subsequently computers.

THE SOUND OF MONEY

At the same time as the typewriter was being pioneered, another American invention would eventually put another machine in every shop and café, bar and restaurant in the world—this machine is the cash register.

If you're old enough (or just a music fan), bring to mind the start of Pink Floyd's classic song "*Money*" from the *Dark Side of the Moon* album. That's right, what you can hear is the unmistakable sound of a cash register, representing the sound of money. James Ritty, a saloon owner in Dayton, Ohio was tired of employees stealing from him, so he invented the cash register in 1879. His idea was simple: every time an employee made a sale they'd enter the amount in a mechanical adding machine, press *total* and a bell would ring announcing the fact that a sale had been made. The manager would be alerted and be able to keep an eye on the transaction. The addition of a cash drawer, that could only be opened when a sale was made or by a key only the manager would have, made the businesses' cash secure from pilfering. James and his brother John, who had helped refine the idea, manufactured and sold "*Ritty's Incorruptible Cashier*" to an eager market.

Ritty's Incorruptible Cashier is also responsible for that common pricing method where goods are priced in odd amounts such as $4.95 or $4.45. You may have thought that a shopkeeper prices something at $4.95 because the customer may psychologically think it's "*$4 and change*" rather than "*$5*," which it practically is. Not so. The $4.95 price means that the customer will almost certainly hand over a $5 bill rather than the exact change and the sales assistant will have to ring the sale up to give back five cents change. The bell will ring and the sale will be recorded. If the price were exactly $5 a crooked sales assistant could simply pocket the $5 bill and hand over the goods and the manager would be none the wiser.

An antique NCR cash register

Over the subsequent years innovations were added such as receipts for customers and sales records for managers, so sales and takings could be reconciled at the end of trading. Ritty's company subsequently became the *National Cash Register Company*, or NCR as it's now known. NCR through its specialization in retail sales equipment is now a major IT service company, as electronic point of sales equipment, as cash registers are now called, have become linked to back-end inventory management and accounting systems.

Meanwhile, in 1886 William S. Burroughs[2] established the *American Arithmometer Company* to manufacture and sell an adding machine. Subsequent machines became increasingly complex; the *Sensimatic* could perform many business accounting tasks semi-automatically. It could store up to 27 separate balances in a ledger and worked with a mechanical calculator called a *Crossfooter*.

---

[2] The name may sound familiar as his grandson, also called William S. Burroughs was the famous writer and Beat Generation poet.

An early Burroughs adding machine

In 1904 the company changed its name to the *Burroughs Adding Machine Company* and was soon the largest such company in the US. Eventually it would be known as just *The Burroughs Corporation*, and by the 1950s it would start manufacturing computers. In 1986 it merged with the *Sperry Corporation* to form *Unisys*, still a major player in corporate information technology. I used to work as a consultant for Unisys.

WEIGHING, COUNTING, TIMING AND TABULATING

The 1880s were a busy time in the US for inventors of gizmos for improving business. In 1885 Julius E. Pitrap patented the computing scale, in 1888 Alexander Dey invented the dial recorder, in 1889 Herman Hollerith patented his electric tabulating machine, and Willard Bundy invented a time clock to record workers' arrival and departure times. All of these inventors and their nascent companies were to merge into a single giant business machine company with the really catchy name of *The Computing Tabulating Recording Company*.

The Computing Tabulating Recording Company grew rapidly with manufacturing facilities across North America, selling machinery including commercial scales, industrial time recorders, tabulators and punched card readers, and even meat and cheese slicers. In 1914 the company poached an executive, Thomas Watson, from NCR. Watson focused the company on customer service, insisting that all the sales staff wore smart dark suits, and expanded operations around the globe. He changed the company name to the *International Business Machine Corporation*.

By 1935, IBM was employed by the US government to keep the employment records for 26 million people for social security, and it was becoming evident that electric tabulating machines, like the ones used for the census, needed to be superseded by something even more efficient. But it wasn't to be the office machine companies that would make the next breakthrough; it would be scientists working for the military on both sides of the Atlantic during World War II. Our quest for the universal machine now becomes top-secret.

Chapter 4

COMPUTERS GO TO WAR

*I*n September 2009 Gordon Brown, the British Prime Minister, issued a statement on Downing Street's official website. The statement read[1]:

"...*Thousands of people have come together to demand justice for Alan Turing and recognition of the appalling way he was treated. While Turing was dealt with under the law of the time and we can't put the clock back, his treatment was of course utterly unfair and I am pleased to have the chance to say how deeply sorry I and we all are for what happened to him... Alan deserves recognition for his contribution to humankind... It is thanks to men and women who were totally committed to fighting fascism, people like Alan Turing, that the horrors of the Holocaust and of total war are part of Europe's history and not Europe's present... Without his outstanding contribution, the history of world war two could well have been very different... The debt of gratitude he is owed makes it all the more horrifying, therefore, that he was treated so inhumanely. In 1952, he was convicted of gross indecency – in effect, tried for being gay... His sentence – and he was faced with the miserable choice of this or prison – was chemical castration by a series of injections of female hormones... So on behalf of the British government, and all those who live freely thanks to Alan's work I am very proud to say: we're sorry, you deserved so much better.*"

What had caused this most unusual government apology?

---

[1] The full text of the apology is in Appendix I

I. Watson, *The Universal Machine*,
DOI 10.1007/978-3-642-28102-0_4,
© Springer-Verlag Berlin Heidelberg 2012

THE TURING MACHINE

Alan Mathison Turing was born in 1912 in Maida Vale London, the son of a British civil servant in India. Like many children of the Empire he was sent at the age of 13 as a boarder to a public school, called Sherborne,[2] in rural Dorset. My elder brother went to Sherborne so I know the school well.

Although he excelled at math, he rarely got good marks as he spent most of his time on advanced study of his own rather than attending to the elementary tasks for which he was being graded. Sherborne believed that the classics were the proper subjects for young gentlemen to study. His work was also always very poorly presented, as if his mind was on other matters and right to the end of his life his notebooks were messy. In addition to mathematics Turing was a keen athlete and particularly enjoyed long distance running.

Alan Turing

---

[2] Note that in England *public* schools are exclusive fee paying *private* schools.

After Sherborne Turing went up to King's College Cambridge, where he studied under some brilliant mathematicians. Although shy and awkward, Turing could hold his own intellectually amongst them. He graduated with first class honors and in 1934, at just 22, was elected a fellow of King's College. The fellowship came with an annual stipend of £300, enough to enable him to follow an academic career.

He started to work on the *Entscheidungsproblem*, or in English, the *decision problem*. Mathematics relies on formal proofs; a complex mathematical statement or formula can be proved to be correct by proving that the formulae upon which it depends are correct, and proving that statements upon which they depend are correct and so on. I'm sure you've seen movies and TV shows were professors cover blackboards with complex formula and finally ecstatically exclaim "*QED!*" QED is an abbreviation from the Latin "*quod erat demonstrandum*" that means, "*what was to be demonstrated.*" The phrase is used by mathematicians and logicians to show that a proof has been completed.

In the 1930s mathematics and the closely related discipline of logic were in a spot of trouble. A brilliant young Austrian, called Gödel, had proven that our mathematical and logical systems were either *complete* (one should be able to prove or disprove every statement) or they were *consistent* (one could not prove a statement that is false or disprove a statement that is true). But, importantly they could not be both complete and consistent.

"*Hold on a minute,*" you're saying, "*I don't understand, and why is this important anyway?*" Well, actually, you do understand, which I can demonstrate through a simple example called the "*liars paradox*".

The liars paradox

This is a popular brainteaser with college kids and stoners, invented in the fourth century BC by a Cretan philosopher, called Epimenides. Let's consider the statement. If the statement is true, then all statements are false, then the statement can't be true, it must be false, in which case it's true, but then it has to be false and so on and so on. You can think about this until it makes your brain hurt, but you can never reach a conclusion.

This paradox was famously used in an episode of *Star Trek* called "*I, Mudd*" in which Harold Mudd destroys an android by forcing it to think about a version of the paradox. As the android goes through the logical consequences of the paradox it's voice gets faster and faster until smoke comes out of its head and finally it explodes. So you see that a logical statement can have a powerful effect if it can't be proved. You might argue that we should just disallow such paradoxes and indeed that is largely what we do. We create logical or mathematical models that are either complete, where every statement can be proved or disproved, or are consistent where true statements can't be disproved and false statements can't be proved.

The mathematical and philosophical establishment, were horrified by Gödel's discovery. It seemed as if the very foundations of their universe had been torn away; could nothing be relied upon? As the mathematician Andre Weil said:

"God exists since mathematics is consistent, and the Devil exists since we cannot prove it."

However, young mathematicians, like Turing, saw the new orthodoxy as fertile territory for research. The *Entscheidungsproblem* relates to completeness and consistency, in that to test for either you have to be able to reach a decision. Put simply, the decision problem is, *"is it possible to reach a decision that a statement is true or false."* So we might say: *"is 10 exactly divisible by 2?"* The answer, of course is *"yes." "Is 10 exactly divisible by 3?"* and the answer is *"no."*

For simple statements the decision problem is easy, you can do it in your head, but for increasingly complex statements it will become harder and we might need to use a formal way of coming to a decision. These formal methods are called *algorithms* – you don't need to be scared of algorithms here's a simple one.

Euclid's Algorithm calculates the greatest common factor (GCF) of two numbers. Consider the following: what is the GCF of 252 and 105? The greatest common factor is the largest whole number that will exactly divide into both numbers and Euclid found a simple, step-by-step, repetitive procedure, an algorithm, for solving it.

First we divide the larger number by the smaller:

$$252 \div 105 = 2 \text{ with a remainder of } 42$$

Now we divide the smaller of the two initial numbers by the remainder.

$$105 \div 42 = 2 \text{ with a remainder of } 21$$

Now we divide the smaller of the two previous numbers by 21.

$$42 \div 21 = 2 \text{ with a remainder of } 0$$

QED. 21 is the greatest common factor of 252 and 105.

We can write this algorithm down more formally as:

a/b gives a remainder of r
b/r gives a remainder of s
r/s gives a remainder of t

...

w/x gives a remainder of y
x/y gives no remainder

y is the GCF of a and b. If the first step produces no remainder, then b (the lesser of the two numbers) is the GCF.

Or even more succinctly in computer code as:

```
function gcf(a, b)
  while b ≠ 0
    t := b
    b := a mod b
    a := t
  return a
```

Using this algorithm we can easily find the greatest common factor between any two numbers. Turing saw that by applying simple algorithms he might be able to satisfy the decision problem. The problem was how to make the algorithms simple enough, yet universal enough, to work on any decision problem, or indeed on all decision problems.

In 1936, whilst studying for a Ph.D. at Princeton in the US, Turing published his breakthrough paper: "*On Computable Numbers, with an application to the Entscheidungsproblem*" in the *Proceedings of the London Mathematical Society*. Turing had an idea; he imagined a hypothetical factory filled with floor upon floor, and rows upon rows of hundreds upon hundreds of computers, each using an algorithm to solve a particular decision problem. Such a factory could in theory solve the decision problem for all problems. He actually at first envisages the computers as women meticulously following algorithms to solve problems like the common factor problem above. Remember, at this time as we've already seen, computers were people who computed.

However, this is where he makes his imaginative leap, realizing that people would be too slow and error prone to realistically tackle the decision problem he imagines a machine that could compute. Turing doesn't imagine a giant mechanical calculating engine like Babbage, nor even an electronic machine, but a purely hypothetical machine now called a "*Turing Machine.*"

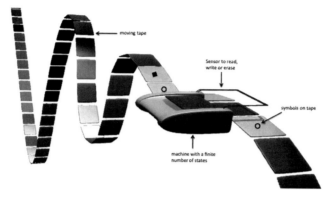

A Turing Machine

A Turing Machine is an imaginary machine that can carry out all sorts of computations on numbers or any type of symbols. Turing saw that if you could write down a set of rules describing a computation (an algorithm) then his machine could accurately carry it out. Thus, a Turing Machine is at the center of the modern theory of computation even though it was invented before the creation of the first computer. A Turing Machine consists of:

- An Input/Output tape,
- The Turing Machine itself, and
- A set of rules.

The tape is a roll of paper that is infinitely long and can be moved forwards and backwards. The tape is divided into cells. The cells can contain symbols.

Above the tape sits a device that can read from, write to, or erase a cell on the tape. The machine can move the tape one cell to the left or right. The machine has an internal state that can be changed. The machine can use the tape as a memory by writing into a cell; since the tape is infinite the machine's memory is infinite. The set of rules, or algorithm is what determines the machine's move at any particular point based on its internal state.

Let's see how a Turing Machine works for a simple example. We want to decide if a string of characters is a palindrome; does the word read the same in either direction. "ABBA" is a simple palindrome; "*Able was I ere I saw Elba*," is a more complex and famous palindrome.

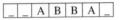

Consider the simple tape above with the word "ABBA" on it; just as with Euclid's Algorithm earlier we can break the search for a palindrome down into a series of simple, repetitive, steps. For our word to have a chance of being a palindrome the first and last letters *must* be the same, if they are not then it cannot be a palindrome. So our Turing Machine can step through the tape starting from the left until it detects the first character "A."

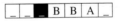

It erases the "A" and then applies a rule looking for an "A" at the end of the word. It advances cell by cell through the tape until it detects the last character.

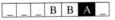

It detects the last character by finding the first blank cell and then moving the tape one cell back to the left.

If the character in the last cell is an "A" our word may be a palindrome. The machine then erases the "A" and moves back to the left until it finds the new first character.

The machine then repeats the steps above, but now on the shorter word "BB". If that is a palindrome, which it is, then our whole word is a palindrome. This algorithm will work for any length palindrome.

_	_	A	B	L	E	_	W	A	S	_	I	_	E	R	E	_	I	_	S	A	W	_	E	L	B	A	_	_	_

The only changes we need to make is first, we need to detect two blank cells to find the end of the phrase instead of just one blank to find the end of a word. Second, we need to have rules for each letter of the alphabet. However, the principle is the same and the algorithm will work as before, comparing the first and last letters, erasing them and then moving on to compare the new first and last letters, until it is determined that the phrase is a palindrome or not.

_	_	**A**	B	L	E	_	W	A	S	_	I	_	E	R	E	_	I	_	S	A	W	_	E	L	B	**A**	_	_	_

_	_	_	**B**	L	E	_	W	A	S	_	I	_	E	R	E	_	I	_	S	A	W	_	E	L	**B**	_	_	_	_

_	_	_	_	**L**	E	_	W	A	S	_	I	_	E	R	E	_	I	_	S	A	W	_	E	**L**	_	_	_	_	_

A Turing Machine couldn't use our wordy description of the algorithm, so we need to translate our algorithm into code for an actual Turing Machine. This code will determine if any word comprised of As & Bs (like ABBA) is a palindrome and print "YES" or "NO" as its answer on the tape.[3]

In our ABBA example above we used just As and Bs, but we might as easily use 1s and 0s. If we do, then we are using binary, which is of course the language that all modern

---

[3] You can see a Turing Machine simulator decide if words are palindromes at http://ironphoenix.org/tril/tm/

computers use. Turing Machines can be created to solve any problem that can be represented by symbols.

Turing had another great insight. He realized that he didn't have to have a separate Turing Machine for each problem; one to detect palindromes another to solve Euclid's Algorithm and so on. You could create a single *universal* Turing Machine that by combining the algorithm and the data on the paper tape could solve *any* problem!

This was Turing's great breakthrough. He had shown that a very simple computational machine could theoretically solve any problem using very simple instructions. *"We may now construct a machine to do the work of this computer,"* he writes towards the end of his paper, and *"It is possible to invent a single machine which can be used to compute any computable sequence."* Such a machine would be a *Universal Turing Machine*.

Consider for a moment: Turing had envisaged a single machine that could do any task that could be programmed into it. This may not seem like a very big idea to you because you've grown up with computers. Your computer is a single machine that can: do word processing, manipulate spreadsheets, send email, surf the web, play games, play music and videos, edit photos, and a host of other things. But, in the 1930s, as we saw in the previous chapter, people used different machines for different tasks. Turing's *universal machine* with its stored program concept was a revolutionary idea – one machine that could do anything! This is why Turing is known as the *"father of computing."*

There is, however, one limitation Our Turing Machine with paper tape runs very slowly, but even if it ran at the speed of light, some problems might never be completed. It may not be possible to come to a decision for every problem. Now there's no point in waiting around for eternity for an answer that may never come. It would be useful to determine in advance if a problem is likely to complete. This is called the *"Halting Problem,"* it has not been solved yet. Turing showed that a general algorithm to solve the halting problem for all possible program-input pairs couldn't exist. The halting problem is *undecidable* over Turing Machines, so he had proven that the

Entscheidungsproblem cannot be solved. Gödel was correct; not all mathematical statements are decidable.

However, fortunately for us we don't need to worry about arcane mathematical theories since many practical and useful problems can be computed in a reasonable time. Thus, at the core of modern computers, are lots of simple tiny machines, performing very simple computations based upon strict concise instructions. In fact it is now almost impossible for us to think of a Turing Machine without thinking of a computer and its hardware and software.

If you search online you can find several examples of simulations of Turing Machines; there is even a YouTube clip of a Turing Machine made from Lego that actually works.

TOTAL WAR

By the start of World War II the navies of the world were dominated by huge battleships with guns so large they could fire shells the size of VW Beetles 20 miles or more. To aim these guns the gunners were dependent on firing tables published in gunnery manuals. The preparation of these tables was laborious as the complex equations were solved by human computers. 100 years after Babbage the US and British navies faced the same old problem of generating tables quickly, but without errors. This problem was replicated for all the land-based artillery used by the huge armies fighting all around the world.

During the war the US military hired every person, usually female math majors, who had the skills to create these firing tables. Don't forget new tables were required for every new type of artillery and shell. Because there actually weren't that many female math graduates the US was not able to produce enough tables, and pieces of artillery were sometimes shipped to the battlefield without firing tables. Without an accurate firing table an artillery piece was virtually useless since it couldn't be aimed accurately. The US military were desperate to remedy this situation and they turned to their best university, Harvard, and their biggest tabulation company, IBM, who between them by 1944 developed the Harvard Mark 1.

A portion of the right side of the Harvard Mark 1

Designed by Howard H. Aiken, the Mark 1 was programmable, like Babbage's Analytical Engine, but it was not an electronic computer. Instead it was an electromechanical hybrid of switches, relays, rotating drive shafts, gears and clutches. The machine was massive, like the Analytical Engine, weighing five tons and standing eight feet tall and 50 feet long. A rotating shaft ran its entire length powered by a five horse-power motor. You can get a sense of its scale by noticing the two typewriters to the right of the picture. The noise the machine made was apparently quite deafening.

It's difficult for us now to understand just how limited this huge electromechanical machine was. Despite its huge size and quarter of a million components, the Mark 1 could only store 72 numbers. Your smartphone today can store billions of numbers. The Mark 1 could add, subtract, multiply and divide numbers of over 20 digits in a matter of seconds, but today your computer can perform calculations in a billionth of a second. Speed was just not possible from a machine with mechanical components – it was faster and more reliable than people though.

In a strange parallel to the Analytical Engine and Ada Lovelace, a woman was also closely involved with the Mark 1. Grace Hopper was one of those rare female math graduates with a PhD from Yale. She was *volunteered* into the Navy in

1943 and quickly ended up on the team developing the Mark 1. Hopper was one of the first programmers of the Mark 1 and is credited with finding the first computer bug. A moth had flown into the Mark 1 and got caught in a relay causing an error – literally a bug! The remains of the bug can still be seen in her team's logbook in the Smithsonian museum. Before performing any computations operators of the Mark 1 would routinely *debug* the machine.

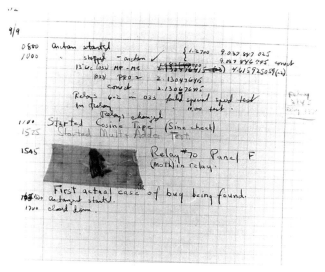

The first computer bug

Hopper later, when working on the UNIVAC computer, invented the first high-level computing language called "*Flow-matic*" that subsequently evolved into COBOL, a language still in use to this day. A high-level computer language looks a bit like English and is easier for people to understand, read and create than the binary that a computer actually understands. A program, called a *compiler,* is needed to translate the high-level language into machine-readable binary. So she also invented the first compiler. After the war Hopper remained in the Navy Reserve and retired as a Rear Admiral when she was 79.

Grace Hopper (January 1984)

Incidentally, Hopper is also credited with coining the phrase: "*It's easier to ask forgiveness than it is to get permission.*" I imagine this is very true in the Navy and it certainly is in university where I work.

CHASING AN ENIGMA

Back across the Atlantic the British were also conscripting mathematicians into the war effort. If you know anything at all about Alan Turing it is probably that during World War II he cracked the German Enigma code and saved hundreds of thousands of lives by bringing the war to a speedier end. Unfortunately like so many popular legends this is not strictly true.

The Enigma code was produced by German electromechanical rotor machines that could be used to encrypt and decrypt secret messages. First invented at the end of World War I they were used by commercial companies and then the German military. The basic Enigma machine was refined and made more complicated during WWII.

An Enigma machine at Bletchley Park

The basic idea of a cipher is to take a piece of ordinary text and scramble it so only the intended recipient can descramble it. The simplest cipher is a *substitution* cipher. For example, take a word and substitute each letter by the letter two positions after it in the alphabet: "*BREAD*" becomes "*DTGCF*". These simple ciphers were used in the Middle Ages for secret messages, but are very easy to crack. To make a cipher more complicated you need to vary the method by which you make the substitution for each letter, but in a way that the recipient can still decipher. This is called *polyalphabetic substitution*.

There is a distinction between *codes* and *ciphers*, though the words are commonly used interchangeably. A *code* is technically where words have different prearranged meanings. I might say: "*Meet me in the park and bring an apple.*" But if you knew in advance that *park* was code for *bank* and *apple* meant *gun* then you'd know to meet me at the bank with a gun. A *cipher* is where the original letters in words have been scrambled.

What the Enigma machine does is to automate polyalphabetic substitution to make a cipher with, at its most complex, approximately 10^{114} possible permutations. Far too many for anyone to crack, or so the Germans thought.

If you look at the photo of the Enigma machine you'll see at the front are a number of electrical wires with plugs and sockets. These wires can be plugged into different sockets to change the machine's basic configuration by swapping pairs of letters. Above the wires is a keyboard, and above the key-board, in the center of the machine a panel of lights that are illuminated letters. Above the lights is a set of small rotors; these can be inserted in different orders and to different settings to further alter the machine's configuration.

To send an encrypted message, the operator chooses which rotors to use for the day, sets the individual rotor's settings and inserts each into the machine in a particular order. He sets the Enigma's plug wirings and the rotor wheels' settings to a predefined initial combination known to him and to the receiver. Then he types the text of the message on the Enigma's keyboard. For each typed letter, a different letter lights up on the panel above the keyboard. The operator's assistant writes down each illuminated letter. When he has finished typing the original message, he will have a seemingly random sequence of letters – the Enigma encrypted message. A radio operator can then transmit the encrypted message by radio using standard Morse code.

The receiving radio operator must write down the received encrypted message, set his Enigma machine to the same pre-defined settings (both electrical and mechanical), and then type the message on the machine's keyboard. As he types his assistant can read the original deciphered text message

An Enigma machine showing the rotors

from the letters illuminated in the panel above the keyboard. The daily settings for the Enigma machine were contained in monthly settings books held by radio operators who were under strict instructions to never let these books fall into enemy hands.

The daily settings from an Enigma settings book

By the end of 1932, the Polish *Cipher Bureau* had developed a laborious method for cracking the Enigma's ciphers. The Poles had good reason to fear the Nazis and before the outbreak of WWII they shared their decryption techniques and equipment with French and British military intelligence. The British understood the Polish techniques for breaking the Enigma cipher and had a working copy of an electromechanical machine designed by the Poles to semi-automate the process called the *Bomba*. However, the Germans had made improvements to the Enigma machine that made it more complicated than the pre-war version. In particular, the Naval Enigma machine used additional rotors, which made it much more complicated to crack. The German Navy used Enigma to send messages to their U-Boat wolf packs in the Atlantic that were decimating the British convoys. It was vital that Enigma was cracked if Britain was not to be starved into submission.

At the outbreak of war Turing went to work at the top-secret *Government Code and Cipher School* housed at Bletchley Park,[4] which I visited recently. If you are ever in the vicinity I can highly recommend the trip; it was fascinating. Many of the photographs in this chapter I took there. During the war this establishment was filled with the best and brightest mathematicians and cryptographers as well as people who excelled at crosswords and all sorts of puzzles. Many people think that Bletchley Park went by the code name "*Station X.*" But it was explained to me there that if you think about it, calling a top-secret base "*Station X*" is a bit daft, because it rather calls attention to it. In fact Bletchley Park was a radio receiving station and it was the 10th in a series of similar stations, hence it's designation *Station X* using the Roman numeral.

Whilst I lived in England I had the privilege of meeting Donald Michie, a colleague of Turing's, who worked at Bletchley Park with him. Donald told me that Turing liked to play chess, but surprisingly for a genius, though a very enthusiastic player, he wasn't very good at the game. He also told me that Turing was such a keen long distance runner he sometimes used to run 40 miles into London, and back, for meetings. Around this time it is recorded that Turing ran a marathon in 2 hours, 46 minutes, and 3 seconds. His time was only 11 minutes slower than the winner of the marathon in the 1948 Olympic Games. Clearly Turing was a competitive long-distance runner.

In collaboration with the mathematician Gordon Welchman, Turing designed a new version of the Polish Bomba, called the "*Bombe,*" to handle the new more complicated version of Enigma.

The Bombe didn't decipher the cipher, but it could search the billions of possible permutations of the Enigma settings and rule out those that were incorrect. This allowed human code-breakers to concentrate on the more probable settings.

---

[4] http://www.bletchleypark.org/

Enigma had one particular weakness: it was impossible for an Enigma machine to code a letter as itself – an "*A*" would never be encrypted as an "*A*." This sounds obvious, even sensible but it gave code-breakers a vital way into the code. This, in combination with other factors such as operator mistakes, and occasional captured hardware and settings books, allowed the cryptographers to be so successful.

A common mistake that was exacerbated by military protocol was to send messages containing the same phrases. This was actually very common as military jargon is standardized and repetitive. One phrase that was often sent was "*no special events*" or "*keine besonderen Ereignisse.*" German units would send this message daily if all was quiet in their sector and there was nothing to report and as any soldier will tell you war can often be quite dull, so this message was sent a lot. Every night for a while one German unit sent the same message "*Beacons lit as ordered*" or "*Feuer brannten wie befohlen.*"

Knowing that a message could contain a known phrase gave the code-breakers an advantage. Another common error was to transmit standardized messages at standard times, such as weather reports to German submarines. If the Allied meteorologists predicted the weather the same as their German counterparts then they might already know the basic content of the message. So a message might read: "*Forecast to 12:00 hours wind south westerly ten knots sea state slight visibility good. HH.*" These standard pieces of expected text were called *cribs*.

If they suspected a crib might be in the transmission they could more easily decipher it, and thus deduce the Enigma machine's settings for the day. Here's how their process worked. They could take the intercepted encrypted German radio message, suspected of containing the crib, and run it through various settings of their Enigma machine and see if decrypted plain German text emerged. If German didn't come out they could change the Enigma settings and repeat the process. Of course there were hundreds of millions of possible Enigma settings and somebody would need to check the output for German each time.

Here's where the flaw in Enigma's process came to help them. They didn't need to check if the possibly decrypted text was German because if any of the letters in the original intercepted message and the possible decrypted message matched (i.e., were identical) then those Enigma settings could not be correct, since Enigma never encodes a letter as itself. Checking to see if letters in the encrypted message and possible decrypted message matched was easier than checking for German language.

This would of course still be a very laborious process if done by hand and also it would be prone to error, but Turing knew the process could be mechanized. A machine could be built that could automate the entire process. In fact, before the war the Poles had built just such a machine, which they called a Bomba. It could simulate several Enigma machine settings simultaneously and evaluate nearly 9,000 different settings an hour. However, since the start of the war the Germans had made the Enigma machine more complicated by the use of the plug board and additional rotors.

Turing's *Bombe* could simulate 30 Enigma machine settings at a time and would rapidly crank through setting after setting until it found a setting where no letters matched in the crib and the encrypted message. An operator (usually a female naval ensign called a WREN) would read the Enigma settings off the Bombe, set up her Enigma simulator to the same setting, type in the encrypted message and if German came out they had the day's Enigma settings. If not the Bombe would be restarted.

The front of the Bombe at Bletchley Park

The photo of the front of the Bombe shows 12 sets of three rotors in three panels. Each vertical set of three rotors represents an individual Enigma machine. The middle panel has an extra set of three rotors on the right. When a Bombe run stops the operator can read the Enigma settings off these rotors.

The front looks complex enough, but step around the back and there is over ten miles of wiring and numerous electromechanical relays. If you visit Bletchley Park you can see a working replica of the Bombe and be taken step by step through the entire process, from the encoding of a message using a real Enigma machine to obtaining the Enigma settings using the Bombe and decoding the message.

Bletchley Park turned military intelligence into an industrial process. 12,000 people worked there in three eight hour

The back of the Bombe at Bletchley Park

shifts 24/7. At any time there might be 4,000 people on duty. Two hundred Bombes were built by the *British Tabulating Machine Company,* which manufactured and sold Hollerith Tabulators under license in the UK. Teams of mechanics serviced the Bombes, for they were unreliable, and further teams of mostly female code-breakers ran them. Once one team had the day's Enigma settings further teams would set about decoding all the day's intercepted German radio traffic.

By the end of the war the Allies were reading all of the German Enigma transmissions, but military intelligence had to be careful how it acted on the information it gained from the German's encrypted messages lest the Germans realize that their communications were no longer secure. This must have caused a terrible moral dilemma for those in authority, as no doubt many lives were sacrificed to ensure that the Allies could continue to read the German's messages. It has even been alleged that Winston Churchill allowed the city of Coventry to be destroyed by the Luftwaffe rather than risk letting the Germans suspect the Allies had cracked Enigma. However, some historians and some who worked at Bletchley refute this allegation. Nonetheless it does serve to illustrate the terrible moral dilemma Churchill would have faced.

A top-secret intelligence group called *Ultra* was established to act upon the intercepted Enigma messages. To cover their success they persuaded German intelligence that their secrets were being leaked by double agents within the German establishment. The Germans spent a lot of time and effort during the war hunting down spies that didn't exist!

Turing's Bombe was not a universal machine, it was a very specific machine completely tailored to the Enigma problem. It was treating each possible Enigma setting as a decision problem and was using a simple algorithm to solve it. After the war all the Bombes were destroyed, because there was no further need for them. Bletchley now has two replicas, one created for the movie *Enigma*, which does not work and a fully working replica built by enthusiasts.

HITLER'S SECRET WRITER

In 1941 the British starting intercepting radio signals that were quite different to the Morse code used to send Enigma messages. They called it "*new music*," and it sounded like the rapid squeal that a fax machine makes; this was a radio signal generated from a telex machine. It was also encrypted, but by a machine the Allies had never seen. This new cipher machine was used on teleprinters by German High Command to communicate with Army Group Commanders. It was initially only used on telephone lines and was considered completely secure. Later as the German forces became stretched out in the field it was also used on wireless transmissions. At Bletchley Park it was codenamed "*Tunny*."

The Germans called the machine their "*Geheimnis Schriftsteller*" or "*secret writer*" and unlike Enigma it was state-of-the-art. Using Enigma was a slow and labor intensive process; typically one person keyed in the original message, another wrote down the cipher text and a third transmitted it via Morse code. The whole process was reversed at the receiving end – only short messages could be sent using Enigma. The new secret writer, or *Lorenz machine* as it was actually called, automated the whole process: a person typed in the message, the Lorenz machine encrypted it and transmitted it via telex and at the receiving end a Lorenz machine automatically decrypted it and printed out the original message. Long messages could be sent with complete security, which was why German High Command used it.

The code-breakers' problem was that, unlike Enigma, they had no way into the code since they'd never seen a Lorenz machine. Fortunately for the Allies on August 30 1941 two German Lorenz operators made a terrible mistake. A 4,500 character long message was transmitted from Athens to Vienna but the receiver in Vienna said the message wasn't properly transmitted. The Athens' operator retransmitted the message using exactly the same Lorenz settings. To make it worse instead of keying exactly the same message the Athens operator used abbreviations in the retransmission so the second message was shorter – only 4,000 characters long. The Allies now had two long similar messages sent using the same Lorenz settings. It was the stroke of luck the code-breakers needed. From this mistake a young mathematician, called Bill Tutte, was able after days of work to find a 41 character repeat in the cipher. Over the next few months Tutte, working with others at Bletchley Park, was able to reverse engineer the Lorenz machine and infer its complete logical structure. Tutte's achievement is still considered a remarkable intellectual feat.

A Lorenz machine at Bletchley Park

After the war the Allies captured some *Lorenz* cipher machines, which were like big Enigma machines with 12 rotors – they were more than twice as complicated as Enigma and they worked exactly as Tutte had inferred. From the logical structure of the Lorenz machine Turing devised a laborious technique, named "*Turingery*," to obtain the Lorenz settings from a cipher text. Max Newman, who ran a group at Bletchley called the *Newmanry*, designed an electromechanical machine nicknamed "*Heath Robinson*," after the cartoonist who drew fabulously intricate contraptions, to automate this process. Unfortunately, the Heath Robinson was slow and unreliable and took up to eight weeks to find the correct Lorenz settings from an intercepted cipher.

The Tunny project required more computing power than Heath Robinson provided; relays just weren't fast enough to sift through all the permutations of the Lorenz machine. This resulted in Bletchley commissioning the *Post Office Research Station* to develop the world's first programmable, digital, electronic computer. It's another common falsehood that Alan Turing designed and even built this machine. This is not true, though his ideas certainly influenced its design.

COLOSSUS

Let's kill another falsehood right away. Sorry guys, the Americans did not build the first electronic computer. ENIAC (*Electronic Numerical Integrator And Computer*) was first used in February 1946 at the University of Pennsylvania. Colossus Mark 1, was working in December 1943. Why then do so many histories state the Americans built the first computer? The reason is simple; Colossus was classified top-secret during the war and remained so until the 1970s! Everyone working with Colossus was bound by the *Official Secrets Act* and would be tried for treason if they spoke or wrote about it. ENIAC was built after the war and was publicly announced and thus entered all the history books as the first computer. Regrettably, because of the secrecy surrounding Colossus its creators did not receive the accolades, awards and fame that they deserved during their lives.

The publicity surrounding ENIAC became so entrenched that for the rest of the twentieth century countless history books, magazines, newspapers, documentaries and even websites continued to spread this falsehood. The Americans did **not** build the first electronic computer – the British did during WWII.

Colossus was the link between Turing's pre-war work on the universal Turing Machine and his post-war work on digital computing. ENIAC was developed in complete ignorance of Colossus, because it was a closely guarded secret, and the British considered ENIAC to be just a big *"number cruncher"* with less sophistication than Colossus.

An RCA triode valve or vacuum tube

An electrical engineer called Tommy Flowers, who worked for the Post Office designing telephone exchanges, saw Newman's Heath Robinson machine at Bletchley and he knew he could build a faster version using electronic valves instead of relays. Colossus Mark 1 had 1,500 valves, sometimes called vacuum tubes, whilst Mark 2 had 2,400 and was four times faster than the Mark 1. Electronic valves operate much quicker than the electromechanical relays in machines, like the Harvard Mark 1 and the Bombe. Relays have to flip a small mechanical switch, whereas valves just move a beam of electrons. The use of state-of-the-art valves was crucial in developing fast digital computers.

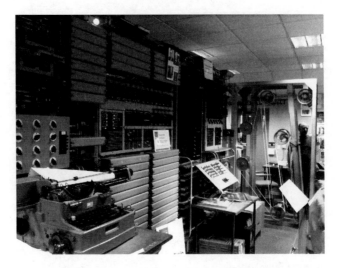

The Colossus Mark 2 computer at Bletchley Park

Colossus optically read a paper tape, which you can see to the right of the photo running around some wheels, and then applied a programmable logical function to every character, counting how often this function returned "*true*". Although machines with many valves were known to have high failure rates, Tommy Flowers knew that valve failures occurred when switching on a machine, in much the same way that light bulbs sometimes pop when you turn them on. Colossus, once turned on, was never powered down unless it malfunctioned. By the end of the war 10 Colossus computers were installed at Bletchley Park. Each computer generated so much heat that the female operators used to bring their laundry into its room to dry and an American visitor, one freezing winter's day, was heard to say that it was, "*the only goddam place in this country that's warm!*"

Colossus was the first electronic digital machine that could be programmed, albeit in very limited ways by modern standards. It was not, though, a Universal Turing Machine, and it was not then realized that what is now known as *Turing completeness* was important. All of the other pioneering computing machines were also not Turing complete. The notion of a computer as a general-purpose universal machine, that is, as more than a calculator devoted to solving hard, but specific problems, would not be seen as important for many years.

By the end of the war the Allies were reading all Enigma transmissions and most of German High Command's Lorenz traffic. Whilst at Bletchley I was told that if in the last days of the war if a German officer had wanted to know the location and status of any particular German unit he'd get the answer quicker by telephoning Bletchley than by asking his own High Command. Historians all seem to agree that the contribution of Turing and all the other code-breakers at Bletchley Park shortened the war by perhaps as much as two years and saved millions of lives. Some historians go even further and suggest that without Turing the Allies would not have won the war at all!

CALCULATING SPACE

The Germans in addition to constantly improving their cryptography machine were also developing computers. It seems though that the Nazis didn't fully understand the importance of computing to the war effort. The first German computer was built by a brilliant lone amateur. Konrad Zuse was a civil engineer who also designed poster advertisements. He worked for Ford in Germany and then in 1935 for a plane factory. This work required him to perform countless complex computations by hand. Like Babbage before him, he started to think about mechanizing the computations. Working alone in his parent's home, completely isolated from mathematicians and university research, he completed the *Z1* in 1938. The Z1 was mechanical, had 30,000 metal components and used 35 mm camera film as punch tape for program instructions. The Z1 never worked because Zuse couldn't engineer the parts to a sufficient precision. The parallels with Babbage continue when in 1989 *Siemens*, the German technology company, sponsored an expensive rebuild of the Z1 even though its original plans had been destroyed during the war.

Zuse was enlisted into the army in 1938 but was eventually given some resources to build the Z2 and Z3, in which he now used relays. The German aviation industry in particular was keen to support his work as they relied on numerous complex calculations to design their increasingly sophisticated planes. All his prototypes were destroyed by Allied air raids and after Germany's defeat there were no resources to continue development.

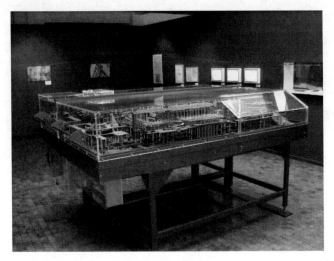

The replica Z1

When Germany recovered economically Zuse was eventually able to start his own computer company, which was bought by Siemens in 1967. Zuse built over 250 computers but is more famous now for a theoretical idea he published in a book called the *Rechnender Raum*. Translated as *Calculating Space,* Zuse proposed that the universe was actually a construct running on a network of computers. This idea is still influencing thinkers today in philosophy, science and entertainment. Recently, the 2010 movie *Tron: Legacy*, about a digital world inside a computer, pays homage to Konrad Zuse – Michael Sheen plays a nightclub owner called *Zuse*. The real Konrad Zuse died in December 1995.

THE TURING TEST

Turing, because of his background as a mathematician, is often considered to be a theoretician and to have had little direct influence on the actual design of computers. This could not be further from the truth. Turing was a skilled electrical engineer and after the war he got a job at the *National Physical Laboratory* where he worked on the design of the ACE (*Automatic Computing Engine*), which was to be one of the first stored-program computers. Turing developed the complete circuit design for ACE and worked hard to make the hardware as simple as possible to improve reliability. He also developed the first software for the ACE and was a keen programmer.

ACE, like all modern computers, stored its program and the data upon which the program operated in the same memory; this is called a *Von Neumann architecture*. Machines like the Harvard Mark 1 had their programs hardwired into them so no program memory was needed. Colossus was programmable, but loading a new program was a laborious task by today's standards. As Turing commented in a report on ACE, "*There will positively be no internal alteration [of the computer] to be made even if we wish suddenly to switch from calculating the energy levels of the neon atom to the enumeration of groups of order 720. It may appear somewhat puzzling that this can be done. How can one expect a machine to do all this multitudinous variety of things? The answer is that we should consider the machine to be doing something quite simple, namely carrying out orders given to it in a standard form which it is able to understand.*"

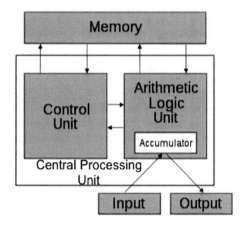

The Von Neumann architecture

Turing eventually became tired of the secrecy of working at government labs and went back to Cambridge for a sabbatical year and then moved to Manchester University where he became Deputy Director of its new *Computing Laboratory.* Here he developed software for the new Manchester Mark 1 computer, an early stored-program computer. Again his work was of a very practical nature including writing: system software to make the machine usable, interface software for a paper tape reader, a random number generator, and what is probably the world's first programming manual – *Programmers' Handbook for the Manchester Electronic Computer Mark II.*[5] In this he lays out his "*Programming Principles: i) Make a Plan. ii) Break the problem down. iii) Do the programming of new subroutines. iv) Programme the main routine.*" This will be still instantly recognizable to any programmers reading this book.

In 1950 he published another theoretical paper that was to have profound and long reaching impact. This would firmly cement Turing as the greatest computer scientist who has ever lived.

"*Computing machinery and intelligence*" was published in *Mind* in October 1950, and as its title suggests Turing was dealing with the subject of machine intelligence, or artificial intelligence as we now call it. In the paper Turing asks what it would for take a machine to be considered "*intelligent*" and he proposed a test for machine intelligence. This test is now known as the *Turing Test*.

In a Turing Test a person sitting at a computer communicates via the keyboard (like instant messaging) with two entities. One is another person and the other is a computer. The person is allowed to ask the two entities any questions they want. If after a set period of time the person cannot tell the machine from the human, then the machine has passed the Turing Test.

---

[5] A facsimile of the manual can be seen on the The Turing Archive for the History of Computing http://www.alanturing.net/turing_archive/archive/m/m01/m01.php

If you think about this test it is a really demonstrating that the machine *appears* to be intelligent. It is not claiming that the machine actually *is* intelligent. You could say that the computer is imitating intelligence. Indeed this is exactly what Turing thought and he wrote: "*Are there imaginable digital computers which would do well in the imitation game?*" Turing believed that this question could be answered positively although nobody could build a computer to even attempt the test in his lifetime.

Not content with inventing the disciplines of computer science and artificial intelligence Turing also turned his remarkable mind to a biological problem – *morphogenesis*. In a paper called *The Chemical Basis of Morphogenis* he described the *Turing hypothesis* of pattern formation, which explains how spots may occur on a leopard's skin for instance. Turing is now considered a founder of bioinformatics.

Alan Turing, 1951

TO DIE PERCHANCE TO SLEEP...

On the June 8 1954 Turing was found dead by his cleaner. He had taken a bite from an apple poisoned with cyanide and the subsequent inquest found he had committed suicide.

During my brief biography of Alan Turing I have omitted a significant detail about him, which though not relevant to his professional career, his remarkable discoveries and contributions, is of great significance to his private life. Turing was homosexual and had been since his school days. We now don't care about people's sexual orientation, but back in the 1950s homosexuality was still illegal in Britain and could result in imprisonment.

In 1952 Turing reported a burglary at his house to the local police. He was able to name a prime suspect, Arnold Murray, a young man he had met a few days before outside a cinema in Manchester. During the police investigation Turing naively told the police that he and Murray had a sexual relationship. Turing and Murray were charged, tried and found guilty of gross indecency.

Turing was given a choice of sentence – imprisonment or chemical castration by estrogen injections. He chose the latter. He also had his security clearance revoked, and though his passport wasn't confiscated, he was denied entry to the US because he now had a criminal conviction.

Nobody really knows why Turing seems to have volunteered to the police details of his illegal relationship with Murray. They were probably suspicious of the relationship between an eminent scientist and a young working class man anyway, but crucially they had no proof. Did Turing believe his position would protect him? Did he want to be caught? We just don't know.

We also don't know why he decided to kill himself, though it is believed that the unnatural hormonal changes to his body may have led to a deep state of depression; he was seeing a psychiatrist before his death. We do know why he chose to eat a poisoned apple. Disney's animated masterpiece *Snow White & the Seven Dwarfs* was Turing's favorite movie and it is believed that he was mimicking the poisoned apple from the story. There is a difference though: Snow White falls into a deep sleep, she doesn't die after eating from the witch's apple. A kiss from a prince awakens her.

There is a popular myth that the Apple logo, an apple with a bite out of it, honors Alan Turing and his poisoned apple. There might also even be a play on words between *bite* and *byte* a computer term. Unfortunately, though I love this myth, Apple historians say there is no truth to it. Stephen Fry the British TV presenter, comedian and friend of Steve Jobs, recalls Jobs saying when asked about the origins of the logo, *"It isn't true, but God we wish it were!"* Whenever I see the Apple logo I remember Turing though, for without his discoveries Apple's products would not exist.

Apple's logo

TURING'S LEGACY

Unfortunately, because of the secrecy surrounding Turing's work during the war, and because of the double taboos surrounding his death – homosexuality and suicide, Turing remained virtually unknown for years. Since 1966 the *Association of Computing Machinery* has named its highest annual award the *Turing Award*; it's considered the computing world's equivalent to a *Nobel Prize*. However, outside of the computer science community almost nobody had heard of Turing. It wasn't until 1983 when the mathematician and author Andrew Hodges published a book called *Alan Turing: The Enigma* that a wider public learnt of Turing. The book was quickly followed by a very successful West End and Broadway play called, *Breaking the Code*. The BBC later broadcast a version of the play on television.

The true history of Turing's great contributions both to Britain's war effort and to the development of the computer has since led to a succession of different honors being conferred on him: a blue plaque was unveiled at his birthplace in London, a main road in Manchester was renamed the *Alan Turing Way* and a bridge the *Alan Turing Bridge*, a statue of him was unveiled at Bletchley Park and another in Manchester with a plaque that reads "*Alan Mathison Turing 1912-1954 Father of Computer Science, Mathematician, Logician, Wartime Codebreaker, Victim of Prejudice.*"

Alan Turing's statue at Bletchley Park

In 1999 *Time Magazine* named him as one of the "*100 Most Important People of the 20th Century*" stating: "*The fact remains that everyone who taps at a keyboard, opening a spreadsheet or a word-processing program, is working on an incarnation of a Turing machine.*"

In 2009 John Graham-Cumming, technologist and author, started a petition asking the British government to apologize posthumously for persecuting Turing for being homosexual. The petition quickly received over 30,000 signatures and the Prime Minister issued a rare official apology.

Turing's name now sits comfortably amongst the great thinkers of our civilization: Socrates, Aristotle, Copernicus, Galileo, Newton, Darwin, Einstein – none of them directly saved as many lives though or could run a competitive marathon.

The subsequent chapters will show how Turing's universal machine has changed our world.

Chapter 5

COMPUTERS AND BIG BUSINESS

*I*t's 1953 and Blair Smith, a senior salesman from IBM, is returning to New York from a business trip to Los Angeles. He's flying with *American Airlines* and in the seat next to him in business class is Cyrus Smith, who happens to be the president of American Airlines. They introduce themselves and realize that they share the same family name. This breaks the ice and they get to talking; it's a long flight back to New York. Each talks about projects that he's been working on.

Blair had been involved with a project for the US Air Force called the *Semi Automatic Ground Environment* (SAGE), which used several huge computers to manage radar intercepts and forward these to airfields so fighter jets can be scrambled; this is the height of the Cold War. People could communicate with this system from US bases all over the world. Perhaps Blair mentioned this project because it involved planes. Cyrus was wrestling with a thorny problem of his own; American Airlines' ticketing system was a manual process that took hours to complete a single booking. The process was creaking at the seams because the number of planes, flights and travel agents was rapidly increasing as the US public embraced the jet age. Blair and Cyrus realized that there were similarities between SAGE and American Airlines' ticketing troubles. A month later Blair sent a proposal to Cyrus suggesting that IBM study the ticketing problem. Blair wasn't about to let such a huge sales opportunity pass him by.

American Airlines' computerized ticketing system was one of the first and most successful commercial application of computers. But, the very first commercial computer application, in 1951, was at a British company called *J. Lyons & Co.*,

I. Watson, *The Universal Machine*,
DOI 10.1007/978-3-642-28102-0_5,
© Springer-Verlag Berlin Heidelberg 2012

famous for its chain of teashops. Before we can tell that story, we have to fill in some background.

ENIAC

After the war, John Mauchly and J. Prepser Eckert, the inventors of ENIAC (the *Electronic Numerical Integrator and Calculator*), continued to work on the development of electronic computers. ENIAC, even by the standard of the machines we have seen so far, was a monster. It completely filled a room 40 feet long; it used more than 18,000 vacuum tubes and weighed 30 tons. Although almost silent when running, unlike the deafening Harvard Mark 1, it generated vast amounts of heat from all those vacuum tubes and had to have serious air conditioning.

Many people, including RCA who manufactured the vacuum tubes, doubted it would ever work since the tubes were notoriously prone to failure; just like light bulbs they would blow when switched on. Eckert is alleged to have paid such careful attention to the circuit design to reduce failures that he even starved some lab rats to find out which electrical cable insulation they were least likely to chew.

A view inside ENIAC

Although the war was over, the military had funded the development of ENIAC, and the first program it ran was to see if it was possible to build a thermonuclear hydrogen bomb (the bombs used on Japan were less powerful fission atomic bombs). ENIAC could perform simple arithmetic tasks in less than three-thousandths of a second. Even so, it took ENIAC six weeks of computation to decide that a hydrogen bomb could be built. This program is still classified "*Top Secret.*"

EDVAC

Programming ENIAC was laborious, requiring physical modifications to countless switches and electrical cords, so Eckert, Mauchly and the mathematician John von Neumann set about designing EDVAC. The *Electronic Discrete Variable Automatic Calculator* could store a program in memory. Note that EDVAC is called a "*calculator*" and not a "*computer*" – computers at this time were still people. Its memory was not like the RAM your computer or smartphone has, but was made from electric pulses turned into mechanical waves in a long cylinder filled with mercury. The pulses moved slowly through the dense liquid, and when they reached the end of the cylinder they were turned back into electrical pulses that could be reintroduced back at the start of the cylinder. In this way the computer could store information as a series of electrical pulses. EDVAC had five and a half kilobytes of memory (5,500 bytes); the computer I'm writing this on has four gigabytes of memory (4,000,000,000 bytes).

EDVAC used 6,000 vacuum tubes, covered almost 500 square feet of floor space and required 30 people to operate it; but it was faster and easier to program than ENIAC. Programs could be read in from magnetic wire and stored in its mercury delay line memory; like Colossus it used Turing's stored program concept. Once again the military had funded its development, and in August 1949 it was delivered to the *Ballistics Research Laboratory,* where it was used for hydrogen bomb development and other military tasks.

There was a dispute over patent rights between Eckert, Mauchly and the University of Pennsylvania, so they left the university and started a computer company called the *Eckert-Mauchly Computer Corporation*. Along with them went most of the scientists and engineers who had worked on ENIAC and EDVAC. They immediately started work on an improved version of EDVAC, which they renamed UNIVAC – the *Universal Automatic Computer*. The US Census Bureau was their first customer; it needed a fast machine to tally up the census returns for the 1950 census.

The UNIVAC I showing the Remington Rand branding

Unfortunately, the new company had a cash flow crisis and lacked wealthy investors. They were well behind on delivering the UNIVAC in time for the census, and the company was sold to *Remington Rand*. The UNIVAC wasn't delivered until March 1951, too late for the census. There were other troubles as well; several leading engineers on the UNIVAC project were accused of being communist sympathizers during the McCarthy era witch-hunts. Mauchly himself was banned from company premises and the company lost its security clearance. He resigned in 1952, but under the terms of his contract he was forbidden to work on any computer projects until 1960.

During the 1950s the UNIVAC was the first mass-produced computer. It got so much media publicity that the word "*UNIVAC*" was synonymous with "*computer*" in the public's mind, just as "*Hoover*" is with "*vacuum cleaner.*" Both EDVAC and UNIVAC were in regular use until the early 1960s, when newer machines made them obsolete. During the 1950s IBM also started making computers and was soon outselling the UNIVACs. IBM became so dominant that by the 1960s the group of eight companies manufacturing computers was known as "*IBM and the seven dwarfs*"

THE MAINFRAME

Electromechanical relays had replaced Babbage's mechanical cogs and gears. Vacuum tubes replaced relays; they were much faster, though unreliable. Now the transistor replaced vacuum tubes. Transistors are semiconducting devices that can amplify and switch electronic signals. In comparison to vacuum tubes, they are small, fast, reliable, and when mass produced cheap.

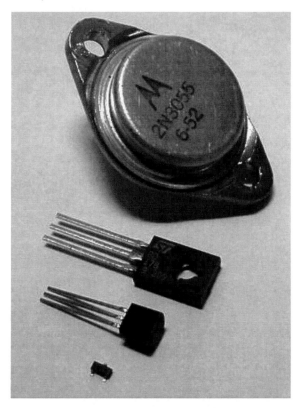

Assorted transistors

The transistor is the basic component of all modern electronics, allowing devices to become smaller, cheaper and more reliable. In the 1960s, small battery-powered portable radios were known as *"transistor radios"* after their key component.

The first experimental computer that used transistors was built at Manchester University in 1953. *Bell Laboratories, Burroughs, Olivetti* and IBM all quickly had transistor computers in development, and by 1957 IBM began delivering the IBM 608, the first solid-state commercial computer. The 608 was soon replaced by the IBM 7090, which became a best seller for IBM.

Two IBM 7090s at NASA in the 1960s

Costing almost three million dollars, most customers preferred to rent a 7090 at a mere $63,500 a month. An early customer was NASA, which used the computers on the Mercury, Gemini and Apollo space programs. Clearly these computers are still large and expensive, and even big organizations like NASA could only afford a couple. Known as *mainframes*, many people could use each of these computers simultaneously, through a process called *time-sharing*, which made them more economic.

In time-sharing many people, perhaps as many as a hundred, are all simultaneously logged in to the mainframe. The mainframe can't process all of the users' requests simultaneously, so it gives some processing time to each user in a round-robin manner. Since the mainframe is fast, it appears to the user as if he or she has sole use of the computer. A common way of communicating with a mainframe was by *teletype*, a form of electric typewriter that transmitted your keystrokes to the mainframe. You typed a line of characters, hit "*return*" and waited for the teletype to noisily print the mainframe's response. Alternatively you could load your program and data in through a paper tape reader, which you can see on the left-hand side of the teletype in the picture. A big advantage of teletypes was that they didn't even need to be near the mainframe; they could be connected via a telephone line. Thus, it was possible to connect to a mainframe by teletype from the other side of the country.

A teletype with a paper tape reader

Mainframes could also be run in *batch mode* – that was when the computer used all of its processing power on just one program. You had to enter your program onto a stack of punch cards using a *card punch* machine. This works just like a typewriter, but instead of typing letters onto paper, it punched holes into cards. Each card contained a single program statement; so your entire program might require a stack of hundreds of punch cards just as Babbage had envisaged over a 100 years before.

An IBM 029 card punch machine

When you'd finished writing your program, you'd load your stack of punch cards into the hopper of a punch card reader machine. Your program would run when the mainframe got round to it. Typically, this would be overnight and in the morning you'd come back in and collect your printout. However, if your program had an error in it, it wouldn't run and your printout would just say there had been an error. You'd then have to laboriously track down the error in your program code, re-punch the incorrect cards and resubmit your program, and probably collect your printout the following day. When I was an undergraduate in the 1980s, the computer center at my university still had teletype machines, keypunch machines and punch card readers that were still in occasional use.

A few years later I was working in a research team with a woman called Margaret. She had strange way of programming, different to everyone else on the project. Everyone else worked on developing his or her programs in small iterative chunks. You'd type a small section of your program into your PC, see if it worked, and if it didn't you'd check for errors and re-enter it, and repeat the process until that part worked. Then you'd move onto the next bit until the whole program was complete. Margaret didn't work like anybody else. She'd read the newspaper and play computer games, sometimes for days. Then she'd suddenly stop what she was doing and write out her entire program on paper in one go. Margaret would then check it over, but usually she'd not find any mistakes. Then she'd enter the whole program into the PC and it would almost always work first time.

I told Margaret that I'd never seen anyone program that way before. She replied that she had learnt to program using punch cards and batch processing. She'd learnt to make her programs perfect before she ever put them onto punch cards. The iterative "error, correct, error, correct" process that we all used would take weeks if you used punch cards and waited overnight for a printout to find your mistakes.

Teletypes were eventually replaced by terminals; these look much like PCs, but are really only *input/output* devices. The keyboard allows character input that can be read on a small monitor or TV screen before being sent to the mainframe. Its response will then also be shown on the screen. Terminals had one big advantage over the teletype; they were quiet, which was important in an office setting. Eventually, terminals would become more sophisticated and even support color displays and graphics, but by then the PC was already taking over. Modern PCs still have *terminal emulator* programs built in, which lets them interface with a mainframe if required.

A televideo ASCII character terminal

Mainframes suited large organizations well; during the day employees could time-share, using first teletypes and then terminals. Employees could make individual inquiries of data, obtain account balances or quantify an item of stock. At night and over the weekends the mainframe could be used to batch process: accounts could be reconciled, monthly statements prepared, sales targets predicted and all manner of complex computations performed.

THE LYONS ELECTRONIC OFFICE

How did an English teashop become the first company to use a computer for business? *J. Lyons and Co.* was not a quaint teashop run by a little old woman. It was a huge restaurant, hotel and food conglomerate, though its nationwide chain of teashops was its best-known business.

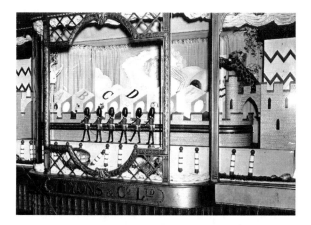

A Lyons window display, circa 1946

After the war the company sent two senior managers to the US to research new business techniques that the Americans were pioneering. They met with one of the developers of ENIAC and were quick to see that a fast, reliable computer could perform many tasks for a large, complex business like Lyons. Of course, if the Colossus project wasn't still top secret they could have learnt about computing much closer to home.

However, the Americans did know that Cambridge University was working on its own version of EDVAC, called the *Electronic Delay Storage Automatic Calculator* (EDSAC). Maurice Wilkes led the team at Cambridge, and when the Lyons managers returned to England they went to meet him. Wilkes wanted funds to speed up the development of EDSAC, and in what seems like a remarkable leap of imagination, the board of directors of Lyons agreed to give him £3,000 and the loan of a Lyons electrical engineer to the project.

Once EDSAC was completed and ran its first program in 1949, Lyons decided to build its own computer improving on EDSAC. This computer was called the *Lyons Electronic Office,* or LEO for short. Again, I'm amazed by the vision of the executives and board members of Lyons. Here's a company that sells tea and bakes cakes building its own computer. If they built nuclear reactors or sold space rockets I'd not be so surprised; even NASA didn't build its own computer in 1949.

The first program to run on LEO was "*bakery valuations*" and by the end of November 1951, LEO did all of the bakery valuations for Lyons. I think we can say that the British had recovered from Babbage's failure to build a calculating machine.

LEO/1 control desk

LEO wasn't a very fast machine; it only had 8¾ kilobytes of mercury delay line memory, but still more than EDIVAC and EDSAC. What makes LEO important is its software, not its hardware. Starting with valuations, LEO was then put to work on the payroll (Lyons had thousands of employees), then inventory management, delivery schedules, costing, invoicing, management reporting and so on. LEO was a universal machine that could be used on all manner of business tasks.

Even with this growing repertoire of applications LEO still had processing time to spare, and in 1956 LEO started doing the payroll for *Ford UK*. Lyons had invented "*out-sourcing*" and eventually would be doing payroll for other companies, and even weather predictions for the *Meteorological Office*. In 1954 Lyons decided to build a new improved version of LEO, and formed a separate company called *LEO Computers Ltd*. This company developed and sold several improved versions of LEO, and *British Telecom* was still using a LEO in 1981. LEO Computers was merged into *International Computers Ltd*

(ICL), one of the companies that later sponsored the Science Museum's Difference Engine build.

ELECTRONIC RECORDING MACHINE-ACCOUNTING

In 1950, Bank of America was having trouble processing its checks. A good clerk could process about 240 checks an hour; that's one every 15 seconds. Assuming an eight hour working day, with no daydreaming, that's just under 2,000 a day – 10,000 a week. That must have been a really boring job. The bank was adding almost 25,000 checking accounts a month and branches were forced to close at two in the afternoon to give them enough time to clear the day's checks. Something had to be done. Bank of America hired the *Stanford Research Institute* (SRI) to investigate automating its book-keeping and check processing.

Over a period of five years (1950–1955) the SRI team revolutionized the way banking was done in two ways. First, they invented account numbers that were unique to each account. Believe it or not, previously accounts were just named after their owner and cross-checks had to be made with other information to ensure two accounts with the same name, such as *"John Doe"* were not confused. This process was known as *"proofing."* Second, they realized that it was important to minimize the number of times clerks entered data, since transcription errors would inevitably occur. Once again we see that the elimination of errors is vital in computerization. Consequently, they didn't want clerks to have to type in a check's account number. They first looked at optical character recognition, but found that minor damage to a check made this unreliable. They considered barcodes, but again damage to a check made reading the barcode unreliable and worse, clerks couldn't read damaged barcodes. So, they invented *magnetic ink character recognition* (MICR).

In MICR, characters are printed in special magnetic ink that can be read by a magnetic reader, a bit like a tape recorder. If the characters are damaged, this has the advantage over barcodes, in that the clerks can easily read the characters.

Misreads are typically less than one in 100,000 and the mechanical check reader could process 10 checks per second.

⑆1234567890⑆ ⑈1234567890⑈ ⑉1234567890⑉ ⑇1234567890⑇

Magnetic ink characters

They invented a special type font, called "*E-13B,*" that improved reading accuracy. If you have a checkbook you can see the magnetic numbers running along the bottom of each check. This highly stylized font became a symbol of modernity and futurism in the 1960s, and you still see it used sometimes in science-fiction movies.

SRI went on to build ERMA, using 8,000 vacuum tubes and a big refrigeration system to keep the 25-ton computer cool. Transistors were still too experimental in 1955, and Bank of America was keen to announce to the world that it used an "*electronic brain*" to manage its customers' accounts. Computers were very popular with the public in the 1950s and considered absolutely cutting-edge technology.

Bank of America subsequently hired *General Electric* to transfer the basic ERMA design to a transistorized computer. Forty new ERMA computers were installed in California in 1959, processing 33,000 checks an hour or 5.5 million a week. Bank of America had solved its check processing bottleneck. It could expand its customer base, offer new types of services, and keep its branches open longer without having to employ more clerks. Tom Morrin of SRI said: "ERMA was the absolute beginning of the mechanization of business." He was almost correct – LEO was.

FLY THE AMERICAN WAY

In the early 1950s, American Airlines' ticket system used a manual process that dated back to the 1920s. Eight employees in Little Rock, Arkansas would receive a ticketing request by phone. One of them would sort through a rotating file, like a Rolodex, with a card for each flight. When a flight was full, the card was marked on the side so a full flight could be quickly identified. The whole process of searching for a flight's card, booking a seat and writing up a ticket took an hour and a half on average, and sometimes much longer. The number of planes and flights was growing rapidly as Americans took to the skies to "*fly the American way.*" Clearly a better process was needed, as this just couldn't keep up with demand.

This is where Blair and Cyrus come back in and have their fateful meeting. A team of IBM analysts and American Airlines staff designed the *Semi-Automated Business Research Environment* (SABRE). The system ran on two IBM 7090 mainframes based in New York and cost a staggering $40 million to build, but it was a huge success. In 1964 SABRE took over all of American's ticketing. It was designed to handle over 8,000 daily transactions. Initially only American Airlines staff used SABRE, but by 1976 travel agents could log in directly using terminals in their offices.

SABRE was spun out into its own company called *Sabre Holdings* and its core functionality is still in use today, connecting tens of thousands of travel agents with more than 400 airlines and almost 100,000 hotels, as well as car-hire, railways, ferries, and all manner of travel and holiday products all over the world. Sabre Holdings is now a global travel technology company, and for example owns the *Travelocity* website. If you've ever booked a flight at a travel agent's office you may have seen them use a SABRE terminal; they're easy to spot since they now look very outdated. They still do the business, though.

By the 1960s every type of large organization, from banks to insurance companies, airlines to hotel chains, manufacturing companies to government agencies, were using mainframe computers for all kinds of business tasks. If they felt they couldn't afford a mainframe themselves, then they could out-source their computing needs to bureaus set up to provide this new service. Only small businesses still did everything on paper and by hand. The mainframes and the software that ran on them had become the universal machines of big business but they had yet to break into the home market.

Chapter 6

DEADHEADS AND PROPELLER HEADS

*T*he year is 1968, and you've been invited to watch a demonstration of a new computer system as part of the *Fall Joint Computer Conference* in the San Francisco Convention Centre. If you are a corporate-type you're wearing a dark suit, button-down shirt and narrow tie (think *Mad Men*). If you're a university-type, you're wearing flairs and a tie-dye shirt or a corduroy jacket with leather patches on the elbows. There aren't many women in the audience and, *shock-horror*, people are smoking…inside! If you use a computer at work you enter your program and data via a noisy teletype or onto punch cards, then wait a few hours, or even over night, for a printout of your results.

The demonstration you watch is pure science-fiction, like a scene from *Star Trek*, which first aired on your TV two years earlier. A man called Doug Engelbart is sitting in front of a keyboard and is using, for the first time in public, a computer mouse. He is in front of a large TV screen monitor that is also projected onto a giant screen in the auditorium. He is interacting with the computer by using the mouse that moves a pointer on the screen to interact with things we would now recognize as windows, but remember you've never seen a graphical user interface before, just punch cards and printouts – computers aren't interactive.

On the roof of the building some longhaired hippy technicians have set up a home made microwave link to the Palo Alto hills beyond San Francisco. These are the same guys who put on psychedelic light shows for the *Grateful Dead* rock band. In the auditorium they are manning a movie camera, TV cameras, lighting, sound and special effects equipment, all coordinated by a man called Bill English. They're using the

I. Watson, *The Universal Machine,*
DOI 10.1007/978-3-642-28102-0_6,
© Springer-Verlag Berlin Heidelberg 2012

microwave link to connect Engelbart's computer in the auditorium to computers at a lab in the Stanford Research Institute (SRI). Another team in SRI are waiting to take part in the demo, and all the various elements of the demo are going to be weaved together on to the big screen in real time. They don't even stop when they have to change reels on the movie camera; it's all done in one take; a bit like the Hitchcock movie *Rope.*

Engelbart shows the audience how he can manipulate data on screen using the mouse and summon up different views of the data. He then shows how he can link to remote computers at SRI and use programs and data on them and then even collaborates with other people in a shared workspace editing documents together. Remember this is 1968, you're still using punch cards and you've never even heard of networked computers. Engelbart then opens a video link and takes part in a video-conference via the computer with people back at SRI. All of this, the screens, windows and videos are shown simultaneously on the big screen in the auditorium.

Your mind is officially blown.
WOW!!!
You've just been shown a glimpse of the future.

A world where computers help people communicate, share and collaborate. You want one of these systems now, but you're going to have to wait 30 years or more for this vision to become real.

Something remarkable happened in a small area around Stanford University, just south of San Francisco in the late 1960s and early 70s – a new industry was born, one that would come to define our age and transform our lives like no other industry in human history. This new industry was not the creation of large business conglomerates, tycoons, or multinational companies, but was invented, almost out of nothing, by college kids, hippies and hobbyists. It came about, not because of the large computer corporations you read about in the previous chapter but sometimes despite of them.

AUGMENTING HUMAN INTELLECT

Before the US entered World War II a man called Vannevar Bush was *Dean of Engineering* at MIT and in 1939 he became chairman of the *National Advisory Committee for Aeronautics*, which we now know as NASA. Overseeing the application of scientific research to the war effort, he had control over the Manhattan Project that developed the atom bomb and projects ranging from radar and sonar to antibiotics. He was therefore not just an administrator, nor an engineer but a wide-ranging and independent thinker. He was particularly interested with how science and technology could augment human capabilities.

In an essay titled: *"As We May Think"* published in the *Atlantic Monthly* in July 1945, Bush put forward the idea of a machine for manipulating a library of microfilm that he called a *memex*. He predicted that, *"Wholly new forms of encyclopedias will appear, ready made with a mesh of associative trails running through them, ready to be dropped into the memex and there amplified."* In September of the same year, *Life* magazine published a shorter version of the essay along with several totally fanciful illustrations of his memex machine.

Vannevar Bush, circa 1944

"It is readily possible to construct a machine which will manipulate premises in accordance with formal logic,

simply by the clever use of relay circuits. Put a set of premises into such a device and turn the crank, and it will readily pass out conclusion after conclusion, all in accordance with logical law"

Now at the same time a young man, Doug Engelbart, was sitting on a crowded troop transport ship waiting anxiously to sail to the Far East to fight the Japanese when news of Japan's complete surrender was announced. However, rather than being allowed to return home straight away he spent the following year in the Philippines. During some downtime, he read Bush's *Life* magazine article. The notion of building a machine to improve or augment a person's memory and problem solving abilities fascinated him and building the memex became his obsession. After his national service he returned to the US and began a career as an electrical engineer, moving from position to position until he finally ended up in California at the recently founded Stanford Research Institute.

STANFORD UNIVERSITY AND SILICON VALLEY

Why is there such a large cluster of high-tech companies in Silicon Valley? Well, the roots of Silicon Valley (it's actually called Santa Clara Valley) go back to well before the invention of the silicon chip and the microprocessor to the start of the twentieth century. The San Francisco Bay area has long been home to the US Navy, which in 1909 built the first radio station in the US. This grew to become the fleet's global radio transmission and reception station. At this time radio was cutting edge technology and several high-tech companies located themselves in the vicinity to service the Navy. In addition to ships, the Navy was also a pioneer of aviation and radar (remember there was no separate air force then) and by World War II aerospace companies, such as Lockheed, were established in the valley.

Silicon Valley showing Mountain View, Sunnyvale and
Palo Alto from left to right

An innovation by Frederick Terman, *Dean of Engineering* at
Stanford University, is credited with laying the foundations of
Silicon Valley. What drove Terman crazy was graduating bright
students who then had to move to the east coast to get jobs
with big technology companies like IBM, GE and AT&T – he
wanted to keep the gifted students nearby. So, he pioneered
the world's first science and technology park in 1951 to act as
a source of employment for his graduates, whom he also
encouraged to start their own companies.

He subsequently personally invested in several of the start-
ups. Amongst his students were William Hewlett and David
Packard, who founded an electronics company you may have
heard of, *Hewlett Packard*, helped by venture capital from
Stanford. The leases in the Stanford Industrial Park were
restricted to high-tech companies in the booming electronics
and aerospace industries. It is worth remembering if you hear
people question the value of universities to society and commerce in particular. Silicon Valley is largely the result of a
university technology park. What a success that idea has
turned out to be. Frederick Terman recalled in his twilight years:

*"When we set out to create a community of technical
scholars in Silicon Valley, there wasn't much here and the
rest of the world looked awfully big. Now a lot of the rest of
the world is here."*

Eventually some of the big east coast names like *Xerox* and
Kodak would have offices in the research park and to this day
an address there is a mark of respectability.

However Stanford's role was not limited to just an industrial park; in 1946 it also founded the *Stanford Research Institute*, now known as *SRI International*. SRI was, and remains, a contract research institute. This model is now common with many universities in the US and internationally. The basic idea is that a university is full of very smart people who are not necessarily very good at marketing their expertise, indeed many aren't even remotely interested in commerce. So you establish a *"research institute"* that people and companies can come to and say, *"I have a problem can you help me solve it."* The institute then identifies the appropriate expertise within the university and sells their skills at a market rate including a reasonable profit for the institute. University academics might not be very interested in business but most love problem solving.

The first job SRI did was to develop a replacement for tallow (that's animal fat) in soap for *Chevron*, which is still used by *Proctor and Gamble*. They then advised the *Disney Corporation* on a possible theme park location in Anaheim, and the *Technicolor Corporation* on a better way of printing film stock. You see the pattern? If you have a problem we know some smart people who can help you, for a fee. SRI now has more than a thousand patents, employs 1,700 people and reported a revenue of $470 million in 2009.

It is commonplace for us to think of academics in ivory towers unaware of popular culture and of industrial scientists and engineers beavering away in their labs ignorant of popular media and mores. This is has always been a stereotype and academics and scientists are as likely to be at a local rock concert, theatre show or sports event as the next guy. Well, perhaps not quite as likely, but you know what I mean. Everyone has obsessions it's just that with academics and scientists their obsessions are likely to be work related. Aren't they lucky – they get paid to play with their obsessions.

This part of northern California in the early 1960s wasn't just famous for its booming high-tech industry. This was the home of the counter-culture. San Francisco was the nearest large city and as hippies left the city for the countryside they moved into the Santa Clara Valley and its environs.

Technicians from SRI were putting on light shows for the Grateful Dead and rumor has it that LSD was kept in fridges at SRI. Visiting be-suited east coast industrialists were horrified to see researchers with bare feet, tie-die shirts, beads and long hair lounging on bean bags in common rooms. They were even more shocked to smell marijuana wafting on the air as they were shown around the facilities at SRI.

These young researchers were not just rebelling for the sake of it, they were exploring different ways of working, collaborating and innovating. LSD wasn't even then considered a recreational drug. It was administered in a strictly controlled fashion to encourage and explore creativity. There was even an institute at Stanford that was researching the beneficial uses of LSD and many students and staff were paid to volunteer for the "*LSD experience.*" In more sinister developments the CIA in a project called MK-ULTRA was also investigating the use of LSD for mind-control.

It's impossible to quantify what effect the counter-culture had on innovations in and around Silicon Valley. What is certain though is that it encouraged and liberated people to think outside the box. No idea, no matter how off-the-wall, could be shot down without being listened to. That just wasn't cool. This spirit of anything goes, coupled with well-financed research departments and a certain amount of good timing helped the innovations pour out of Silicon Valley.

What I'm trying to paint here for you is a picture of a place that was unique in bringing together a group of extremely bright and talented people into an environment that supported, nurtured and valued their creativity. This fired the explosion of technological innovation that was to come. In many ways it is this slightly off-beat, anarchic, creatively and intellectually free environment that characterizes Silicon Valley and is still emulated by companies today. For example *Google* is famous for its free canteens, beanbags, beach volleyball courts, and its 20% rule that allows its engineers to spend a fifth of their time on projects entirely of their own choosing. This non-hierarchical technological sandbox approach was what created Silicon Valley. All that needed to be added were the entrepreneurs and the start-up culture for fortunes to be made along with an industry of global importance.

In the early 1960s Doug Engelbart took up a position at SRI, still with his obsession of augmenting the human intellect. He had fallen on his feet and was able to establish the *Augmentation Research Centre* (ARC) to explore and perhaps fulfill his dream.

Basically, at this time computers were large expensive machines that many users had to share at once. Time-sharing was the prevalent way of using computers The idea is that each user gets a slice of the computer's processing time because the computer's processor isn't occupied or busy all of the time. Therefore, it makes sense to share it amongst a group of users. Think about your own computer at home or at work. How often is it sitting idle doing nothing other than perhaps moving the screen saver or playing music in the background?

However, this view of time-sharing computers was at odds with the views held by many people in the Bay Area, within SRI and beyond. These people wanted a computer all of their own. They wanted the entire computer's processing power dedicated just to their needs. For example, they wanted to drive graphical computer terminals and keyboard inputs rather than rely on punch cards and printouts. At many computer labs in the US people, mostly young, would come to work, or play, during the night when time-sharing computers were largely idle. They could run their programs exclusively until people starting turning up to work in the morning. It is here that we got the nocturnal image of the computer geek or hacker, surrounded by pizza boxes and cola cans. Being nocturnal was a necessity if you wanted sole use of a DEC PDP-10 computer back in 1969.

Partly as a consequence of this people wanted their own personal computers. But since computers cost hundreds of thousands of dollars and had tens of thousands of expensive components this wasn't feasible. However, like so much else in the 1960s, *"the times they are a-changing."*

Doug Engelbart wasn't a nocturnal geek but he knew that if the computer was to help augment human intelligence we had to be able to communicate with it in more natural ways; punch cards and printouts were a dreadful form of interaction.

"Doug Engelbart argued for the co-evolution of technology and human capabilities. An example of such co-evolution was the notion that a computer-mediated communication system that encouraged greater flexibility and interconnectivity would promote new office environments better supportive of collaborative work."[1]

Under his leadership the ARC developed many of the components we recognize in computer interfaces today: bit mapped graphical displays, hypertext (a direct implementation of Vannevar Bush's memex and the precursor of the Web), the computer mouse and even a collaborative working environment, called the *oN-Line System,* that linked people and their computers together into a shared work space.

The first computer mouse, held by Doug Engelbart

For example, In 1963 Doug Engelbart and Bill English, the chief engineer at ARC, built the first computer mouse out of a

---

[1] http://sloan.stanford.edu/mousesite/Culture.html

small wooden box and horizontal and vertical track-wheels. It was called a *mouse* because the wire connecting it to the computer was thought to resemble a mouse's tail. The mouse and the graphical display to which it was connected would go on to be used at Xerox PARC, Apple and eventually by Microsoft in Windows.

All these innovations were finally demonstrated to the public, in the demo described at the start of this chapter on December 9 1968. So mind-blowing was Engelbart's demonstration that it has since become known as, "*The Mother of All Demos*" and entered computing folklore.

SAIL

SRI wasn't the only place started by Stanford that was innovating in the use of computers. SAIL, or the *Stanford Artificial Intelligence Lab*, operated out of a futuristic building in the foothills of the Santa Cruz Mountains overlooking Stanford. SAIL was headed by a brilliant mathematician, John McCarthy, who also had a vision of the future, but a radically different one from Engelbart's.

McCarthy founded SAIL in 1963, after leaving MIT in Boston. He coined the now famous term, "*artificial intelligence*" in 1955 and invented the computer language LISP that is still used by AI researchers. In a way the term artificial intelligence gives us an insight into McCarthy's vision. You see, he wanted to create intelligent entities separate from people; intelligent machines that could solve problems and perform tasks independently of us. Engelbart's vision was to "*augment intelligence*" hence the name of his "*Augmentation Research Center*".

Put simply Engelbart wanted computers to help us solve problems and complete tasks better, while McCarthy wanted computers to solve the problems and complete the tasks themselves. Engelbart's vision is closer to that of the universal machine, since he wanted to provide us with a tool that would help us complete *any* task better. He chose the task of developing a software tool to help develop software tools as his example problem. But in principal his system was designed to help us complete *any* complex task better, it would *augment* our intelligence.

Consider the classic AI problem of chess playing. McCarthy's approach would be to build a chess-playing computer. In 1997 IBM built a massively powerful computer, called *Deep Blue*, to challenge the reigning world champion, Gary Kasparov; Deep Blue eventually won 3½ games to 2½.

Engelbart's approach would be to create a computer system that would help you play better chess, perhaps by allowing you to find opening gambits and advise on other strategy. IBM's Deep Blue could only play chess – some cynics claim it could only play chess against Gary Kasparov – but a chess augmentation system might be able to help you play *any* board game better. Engelbart was thinking in a *universal* way. His computers could help you with any problem or task. But, of course, he had to build them first.

There was another major difference between McCarthy and Engelbart. McCarthy was an advocate of time-share computing. McCarthy believed that in the future computing power would be sold like electricity by utility companies and rather than buying your own computer you'd just buy the processing power that you wanted, when you needed it.

Now of course McCarthy's vision of the future is turning out to be correct with "*cloud computing,*" which we'll talk more about later. But, back in the 1960s there just wasn't enough processing power to even keep the handful of researchers at SRI happy. Engelbart and his colleagues knew that the only solution to their highly interactive and graphical vision of the computer was dedicated processing power, a computer for each user – or in other words personal computers. Thus, the need for personal computers was born, partly out of a desire to create a universal machine to augment our intelligence.

PLAYING IN THE PARC

As the 1960s came to end some of the funding for SRI and the Augmentation Research Center (mostly from the US military) started to dry up. Fortunately for many of its employees there was a new game in town. *Xerox*, the photocopier company, had just opened a new research center down the road, called *Xerox PARC* (short for *Palo Alto Research Centre*). Xerox had made a lot of money from photocopiers and were so synonymous with the activity that people would say, "*can you give me a xerox of that please*", or "*I was xeroxing*," yet despite their healthy finances senior management were worried. What if the predictions of the "*paperless office*" came true? Nobody would need to xerox anymore and computer companies like IBM would make all the money.

So Xerox bought a west coast computer company called *Scientific Data Systems* and they decided to take the initiative and look into what an office of the future, full of computers, might look like. In a brilliant decision they decided to locate their new research center in the Stanford Industrial Park so it would be near other hi-tech companies and Stanford and not place it near head office in Rochester, New York. On the west coast they hoped old-school thinking would not influence it. They weren't to be disappointed.

Can you imagine a single company that invented all of the following:

• The Graphical User Interface, you know windows and icons and pointers,

• Bit-mapped graphics,

• The WYSIWYG word processor, think Microsoft Word,

• Postscript, most printers use it to print documents,

• The Ethernet, your workplace, school or college uses this to network computers,

- Object-oriented programming, you've heard of Java or .Net perhaps, and finally,

- The personal computer!!!

Surely, if this company existed it would rule the IT world, bigger than IBM, Microsoft and Apple. Xerox *did* invent all of the above! Well perhaps they didn't actually invent the personal computer but they sure as hell built one before everyone else. So if Xerox did all this why don't they rule the world?

That is a very interesting story. Xerox wanted its new PARC to be away from the influence of the suits at head-office and for it to be a radical, freewheeling, innovative sort of place where good ideas good germinate and blossom. A brilliant leader, Bob Taylor ran things in the new *Computer Science Lab* at PARC. He knew that if they were to do great things they needed great people and he set about hiring the brightest and the best from around the country. He was helped by the fact that many staff from SRI simply moved over en-mass. At this time it was believed that Taylor employed 50% of the computer scientists in the US.

Although head office took a hands-off approach they did lay down one rule. Xerox had recently bought Scientific Data Systems and PARC had to use their Sigma mainframe computer. This irritated the computer scientists at PARC who believed the Sigma was an outdated dog of a computer; they wanted to use the *Digital Equipment Corporation* PDP-10. This was a de facto standard amongst the US research community and was a generation ahead of the Sigma. In fact not only was Sigma outdated, Scientific Data Systems never made a cent for Xerox and its purchase eventually lost them hundreds of millions of dollars.

So when Taylor tried to place an order for a DEC PDP-10 people back at Xerox head office went nuts. DEC was the Sigma's big rival so what sort of message where they sending out buying their competitor's product. It would be as if everyone at Microsoft Research all used Macs! PARC was forbidden to buy the PDP-10.

The guys at PARC were smart, "*OK*," they thought, "*we can't buy a PDP-10 but nobody said we couldn't build our own copy of a PDP-10! Moreover, perhaps we could even build it better.*" Sure they had to buy all sorts of expensive and exotic components but hey, they were a research lab, and the bean-counters at head office didn't know what the parts were for anyway. Just so long as they didn't place any orders with DEC they were fine. The project became known as the *Multiple Access Xerox Computer* or MACX. In fact even its name was a sly dig at head office since Max Palevsky was the man who had sold his computer company and the Sigma to Xerox and walked away into a golden retirement with millions.

Head office never forgot this issue and relations between west and east coast were strained right from the beginning and never really recovered. But as far as PARC were concerned building MACX was a godsend. It brought them together as a close-knit team, including hardware and software specialists, and it forced them to consider what they really wanted from their computers. This tight integration between hardware and software specialists became key to PARC's innovative success and would later be seen again in Apple. Alan Kay, who worked at PARC and later became an *Apple Fellow* famously said, "*People who are really serious about software should make their own hardware.*"

INVENTING THE FUTURE

After building MACX they turned their attention to designing the office computer of the future. Today their vision seems commonplace but then it was science-fiction. They envisaged an office where each worker had a personal computer. The computers would be able to communicate with one another (networked we'd call it) and be able to share documents. There would be a central printer that would print documents to a high professionally printed quality. Moreover, the computers would be simple to operate and intuitive and documents on screen would look exactly as printed, enabling them to be proofed before printing.

This all sounds easy to us now, but let's break this vision down and see what the engineers at Xerox had to do in order to make this real. They had to invent:

1. The personal computer, so each office worker could have one rather than time-sharing a computer like the PDP-10 or MACX.
2. An operating system and programs that were easy and intuitive to use. Office workers couldn't be expected to use the command line interfaces used at the time by programmers; so they had to invent a graphical user interface like Windows.
3. "*What you see is what you get*" word processing (WISWYG) like we see today in MS Word, so documents could be proofed before printing
4. Bitmapped graphics so these documents could display on a computer screen,
5. Computer networking so the computers and printers could talk to one another.
6. High quality printers (in fact they invented the laser printer).
7. A document description language so the document could be sent from a computer to the printer and be understood by the printer's software.

A Xerox Alto computer

And do you know what? They invented each of these within a few short years. As you can see from this photo of the *Xerox Alto* it had a screen that was designed to look like a piece of paper in portrait orientation, there was the now familiar mouse, and while bulky by today's standards it was very personal for the 1970s.

The Alto was only ever used by Xerox and some collaborating universities and research centers. It was later commercialized as the *Xerox Star*. This system was not intended for the likes of you and I, but was marketed as the Xerox "*personal office system*" connecting workstations, file servers and high quality printers via Ethernet. A single workstation sold for $16,000, and a whole system for your office might cost $200,000. You should consider that in the late

1970s a secretary's annual salary was around $12,000. This was going to be a hard sell.

By the day's standards these machines were powerful, but they were still slow. This was the bleeding edge of office technology and simple operations like saving a file took minutes and system crashes were common, and could take hours to recover from. Finally, although Xerox is a household name, then as now it was famous for copiers, not computer systems. Large companies bought their computers from IBM not Xerox.

Xerox struggled on with the Star and its software through various systems and upgrades during the 1980s and finally admitted defeat in 1989. It had the vision, it had the smarts, it had been on top of the mountain first but it just couldn't sell its dream. The inability of Xerox to capitalize on its inventions at PARC has become stuff of business school legend. Xerox can claim to have invented the PC and all the infrastructure required to use them in small businesses yet it failed to make a dollar from any of them! How could this be? Obviously it's management incompetence, a boardroom full of dinosaurs.

Well as they say hindsight is a wonderful thing. Obviously now we can all see that they should have had a monopoly of the office computing market for decades and probably the home PC market as well. But let's cut Xerox some slack here and think about it without the benefit of 20:20 hindsight.

Firstly, as we know there was poor communication between Xerox PARC on the west coast and head office back east. In fact you could say the relationship was dysfunctional. Clearly the problems that caused can be squarely laid at the feet of senior management. But the bigger picture of why they just didn't see the grand vision of an office of (relatively) inexpensive computers linked by a network and able to create, share and print documents is easier to understand.

Xerox was a photocopier company. But Xerox didn't make its money selling copy machines. Its sales force sold leases for copiers and made their money (and commissions) from the ongoing leases, consumables and maintenance of the copiers. So firstly, they just didn't get the idea of "office computers," since back then all computers were huge back office machines used to run payrolls or do engineering calculations. Secondly, salesmen had no interest in selling an idea that might result in fewer leased copier machines. The one thing Xerox salesmen did not want to see was a paperless office, not now, not ever.

Really you could argue that the inventions at Xerox in the 60s and early 70s were just too ahead of the times. Moreover there was no champion at Xerox head office to drive the grand vision of transforming the company from being the dominant copier company to becoming the dominant office computing company.

Here's a personal story that for me puts this into context. Back in the mid 1990s I was a young researcher and I was asked by a professor in my department to purchase him a PC for his office. I was to advise his secretary who would make the purchase for him. So I went to talk to her to find out what sort of computing requirements he had. She explained that each morning she printed out his emails and took them in to him along with the morning's letters. He would read them and later in the day she would go into his office and he would dictate replies while she took shorthand. She would then go back to her office and type up the replies to the letters and emails (she had a PC on her desk) and then she would send them all, print off the replies and file them with the original letters or emails in a filing cabinet. As far as she knew he had no need of a computer.

Now remember this was about 1995 and he was a well-educated professor of Engineering. Yet still, he had a secretary, he couldn't type; in fact I'm sure he viewed typing as "*woman's work*" and although email was used in university he still lived in a paper-based world unchanged since Victorian times. Emails and their replies were anachronistically printed, date stamped with a rubber stamp and filed away in cabinets for future reference. He still hadn't got the vision of office computing in 1995 so what chance did Xerox executives have 20 years earlier.

Knowing he wasn't really going to use the computer, but just wanted it to sit on a desk at the back of his office to look impressive to visitors, I ordered him a cheap 486SX PC. The geeks amongst you may know that these were based on Intel

486DX chips that had defective floating-point chips and consequently sold at bargain prices. I told his secretary that if he ever asked what the "*SX*" stood for she should say confidently, "*Super eXecutive*". He was later observed guiltily using his new computer to play solitaire.

In December 1979 Steve Jobs from Apple visited Xerox PARC twice. These visits have become stuff of computing myth, legend and fable. Like all such stories it is now hard to know what really happened and to separate fact from fiction. So let's start with the popular version of events.

In popular myth Steve Jobs visited PARC, perhaps even without permission, and saw the computer mouse and the Xerox Alto with its revolutionary graphical interface and rushed back to Apple knowing this was the future of personal computers. He and colleagues back at Apple then reimplemented what Xerox had done and in 1984 the Macintosh was released and the rest as they say is history.

It's a great story, clean and simple, with a shining white knight, Steve Jobs, who liberates our technology in a daring raid from the big villain Xerox, a company too dumb to know what it had. Except, like many such stories, it's not actually true. The first thing that may surprise you is that the *Xerox Development Corporation*, a venture capital arm of Xerox, had invested over a million dollars in the start-up Apple. So Steve Jobs was actually invited to visit PARC as part of this funding deal. Xerox wanted Apple to look at their inventions, and Apple wanted Xerox's help.

However, we're getting ahead of ourselves and we need to back up to see how Apple came to be in the first place. It's one of the more remarkable tales in our story of the universal machine.

Chapter 7

THE COMPUTER GETS PERSONAL

*I*t's January 22 1984 and you are watching the Super Bowl on TV. A commercial break is taking place during a stoppage in play. Commercial spots in the Super Bowl are the prime advertising slots of the year; when carmakers unveil their new models. A commercial starts; it's very high quality, more like a movie trailer than a normal TV commercial.

A line of grey-faced men with hunched shoulders and down-cast faces are trudging down a corridor towards a dimly lit hall where rows upon rows of seated grey-suited men stare at a large screen. A man's face in tight close up is lecturing them on their uniformity, superiority and invincibility. Meanwhile, hel-meted riot police chase an attractive young woman into the auditorium. Her red shorts are the only splash of color amongst the drab dreariness. She stops and hurls a large hammer at the screen that shatters and explodes; a voice-over says:

"On January 24th, Apple Computer will introduce Macin-tosh. And you'll see why 1984 won't be like "1984."

You've never seen such an enigmatic commercial before. It doesn't even tell you what a Macintosh is. You'll never see this commercial again either, as it only aired once. The next day, back at work, everyone's talking about it, what's a Macintosh?

I. Watson, *The Universal Machine*,
DOI 10.1007/978-3-642-28102-0_7,
© Springer-Verlag Berlin Heidelberg 2012

WOZ

Everyone has heard of Steve Jobs of Apple, he was one of the most famous business leaders in the world, but you may not have heard or know much about the man who co-founded Apple with Jobs and actually designed the original Apple computers. His name is Steve Wozniak, or as he prefers to be called "Woz."

Woz was a very unusual child, but once again in this story he's a product of the unique environment of Silicon Valley. His story just could not have happened at any other place on Earth. His father was an electrical engineer working in the missile program at Lockheed. His work was top-secret and he never discussed details of it with his family, except once, he showed interest in TV coverage of a Polaris nuclear missile being launched from a submarine. Although he didn't bring his work home he did bring home his love of electronics that he shared with Woz from a very early age, and I mean early. By the age of four Woz relates that he was familiar with electronic components like resistors and capacitors and knew what they were and how they worked.

I think it's fair to say that Woz was a child prodigy, but instead of excelling at the violin or chess it was science and electronics in particular. By the sixth grade, his IQ was over 200! His father taught him electrical engineering from the ground up, from electrons through to complex circuits like the brand new, state-of-the-art, transistor. He also instilled in his son a firm belief that engineering was of the highest importance. Someone who could make electrical devices that do something good for people could move society to a higher level. This belief would be at the heart of Apple.

Woz was able to see beauty in an electrical circuit and to recognize that an efficient design was better than an inefficient one. He also knew that whatever the circuit was part of it had to be usable; otherwise what was the point? Here's a quote from his autobiography *iWoz*:

"I wanted to put chips together like an artist, better than anyone else could and in a way that would be the absolute most usable by humans. That was my goal when I built the first computer, the one that later became the Apple I. It was the first computer to use a keyboard so you could type onto it, and the first to use a screen you could look at. The idea of usable technology was something that was kind of born in my head as a kid."

Woz built his own crystal radio set when he was only six years old and he got his short-wave "ham" radio license when he was just ten. He was the youngest license holder in California. He also built his own transmitter and receiver from kits his parents gave him for Christmas. They were clearly keen to indulge and encourage his talents since the kits cost several thousand dollars; that's a lot of money for a Christmas present, even by today's standards, let alone in 1960!

But the thing was, Woz really wasn't that unusual in Silicon Valley, all of his friends' fathers worked in the high-tech industries clustered together in the valley and their homes, garages and toolsheds were full of electrical components, wires and circuits. Woz and his friends, who called themselves the *Electronics Kids*, grew up building electrical gadgets like walkie-talkies and radios. At about the age of 11 they wired all their houses together with an intercom system created from scavenged components and connected with hundreds of feet of telephone cable given to them by a telephone company man from the back of his van. The intercom had lights and a buzzer to announce incoming calls. The buzzer could be switched off at night so as not to wake the kids' parents. So you see Woz wasn't that unusual for Silicon Valley, just perhaps more gifted than most.

As he moved up through school one of Woz's great loves was competing in science projects and fairs. He went to great efforts and was soon producing amazing electronic devices that were beyond the understanding of even some of the teachers. The US Air Force gave him its top award for an electronics project for the Bay Area Science Fair, when he was just in the eighth grade even though the fair went up to twelfth grade (that's children four years older than him).

When Woz was about ten he came across an article about the ENIAC computer in an engineering journal his father had left lying around at home. Woz at that young age had discovered his dream. He wanted a computer of his own. Fulfilling that dream would result in you and I being able to have our own computers.

Besides his precocious talent for electronics Woz had one other character trait that stays with him to this day—a love of playing pranks. Not simple, stink bombs in the classroom or drawing pins on seats, but complex, elaborate and involved pranks. Once he made a fake time-bomb out of a metronome, some wires and batteries made to look like sticks of dynamite. He hid his fake bomb is a school locker and you can imagine the result. He ended up spending a night in juvenile hall for that stunt. Later, when at college, he built a miniature TV signal jammer that he hid inside a ballpoint pen. He'd turn it on to interfere with the TV picture in the common room and then when a student came to fix the picture he'd play with them making the picture scramble and go back to normal as they tried to tune the TV in. He particularly loved jamming the signal just as a touchdown was about to be scored during a football game. He loved annoying the jocks.

I guess that nowadays we'd call Woz a geek or a nerd, but back in the 60s as he moved up through high school he was just a bit odd. Whilst his peers had moved on from being Electronics Kids to dating girls and going to parties Woz was redesigning minicomputers in his bedroom.

Steve Wozniak in 1983

Minicomputers were the latest thing. Using the newly invented microchips and transistors rather than the valves of computers like Colossus and ENIAC they were the size of small fridges rather than huge roomfuls of equipment. Now Woz didn't own a minicomputer and nor was he ever likely to, but he enjoyed redesigning them using fewer, newer and better components. He was always trying to minimize the number of chips in the computer and it became a secret passion of his. Alone in his bedroom he'd refine his designs over and over until his designs were better than the ones coming out of the new computer companies.

No, he didn't have a girlfriend.

PHREAKING, JOKES AND JOBS

When he graduated high school there was no doubt what he wanted to do—electrical engineering and maybe computing. He initially attended the University of Colorado, which, unusually for then, had a course in computing. Whilst there he pulled a prank, writing a program that caused the expensive printers to print mile after mile of numbers. After a year he returned to California and went to Berkeley; Woz never graduated college. Instead he and his new friend Steve Jobs were to start a company they called Apple.

As a hobby project Woz had designed and built his own computer from scavenged microchips with 256 bytes of RAM. It couldn't do anything, just flash some lights or beep, but it was a computer and he'd built it! Woz showed it to a school friend, who suggested he meet a guy called Steve Jobs, who also liked electronics and pranks. Now you may have read that the two Steves were school friends. This isn't strictly true, although they did attend the same school, Steve Jobs is four years younger than Woz, and as you know, four years is like a lifetime at school, but less of a difference when you're older.

The two Steves complemented each other well—Woz was a technical genius but Jobs had the gift for selling an idea and could talk his way into anywhere. He scored free chips for Woz just by calling sales reps. As Woz commented, "*What one of us found difficult, the other often accomplished pretty handily.*"

The first thing they built and sold was a blue box for *phone phreaking*. Phone phreaking was a technique for taking control of telephone company networks, chiefly so you can make a call without being billed. When telephones were new and few people had them, to place a call you'd dial the operator and asked to be connected to a particular number. The operator would then physically make the connection by inserting plugs into sockets. You've seen operators (usually young women wearing headsets) in movies and on the TV doing this.

A blue box built by Woz on display at the Computer History Museum

As the telephone networks grew this was no longer feasible and the system was automated, initially electromechanically with relays and then digitally. You know those sounds you hear when you dial a telephone number, each with a different pitch. These are listened for by the telephone exchange and drive the placement of calls. What phone phreakers discovered was that by making specific tones into a telephone (some people with perfect pitch could even whistle the tones) they could seize control of lines from free exchanges, like 0800 numbers, and then they could direct the exchange to place calls toll-free.

Obviously, phone phreaking was not viewed favorably by the telephone companies; it was theft. Since the phreakers often made interstate or even international calls, it was also a Federal offence and would be investigated by the FBI. Nonetheless, there was a thriving, though naturally secretive, community of phreakers. Woz built a blue box that generated the necessary tones for phreaking digitally and thus reliably. Once he called the Vatican in Italy and asked to speak to the Pope, claiming he was Henry Kissinger. The people in the Vatican said that if he called back in a while they'd connect him to the

Pope. So Woz called back later to be told, "*You're not Henry Kissinger! We called the White House and Henry doesn't want to speak to the Pope. Who are you?*" Woz hung up.

Steve Jobs, who went by the phreaking pseudonym of *Oaf Tobar* and Woz, a.k.a. *Berkeley Blue*, made and sold blue boxes. This was a very clandestine affair since both local police and the Feds would have been happy to arrest them. So sales were arranged by word of mouth with college frat boys and various counter culture groups. They even rubbed up against the local criminal underworld and were once robbed of a blue box at gunpoint! Clearly it would be hard to build the sales of an illegal device into a profitable corporation. But the two Steves had learnt they were a good team.

Around this time Woz wrecked his car and he dropped out of Berkeley to get a job to afford a replacement. He landed what he considered his dream job at *Hewlett-Packard*. Unlike many technology companies, Hewlett-Packard wasn't totally run by marketing people; it cherished its engineers. Woz was to help design scientific calculators at HP. Steve Jobs also briefly worked at HP, as a summer intern, before working for a new computer games company called Atari.

Now HP was exactly the sort of company that Frederick Terman, the Dean of Engineering at Stanford University, had wanted to spring out of his industrial park. Bill Hewlett and Dave Packard were both Stanford graduates; they founded Hewlett-Packard in their garage in 1939 and grew the company with investment from Stanford. Woz loved HP, a company run for and by engineers, whose customers were engineers. However, he did have time for personal projects on the side.

The one Woz claims to be most proud of is the first Dial-a-Joke phone service in the Bay Area. The jokes were all politically incorrect Polish jokes, you know the type: "*Did you hear about the Polish Admiral who wanted to be buried at sea when he died? Five sailors drowned digging his grave!*"

The service was very popular but a local Polish community group complained that the jokes were derogatory. Woz asked them if they would prefer if he replaced the Polish jokes with Italian ones. They thought that was a great idea! Despite it's popularity Woz had no business model, he was renting expensive answer-phone equipment but nobody would donate money for his poor taste jokes. The telephone company wouldn't invent premium, paid for, service numbers for years.

However, another business opportunity came along again involving Steve Jobs. Now working for Atari, Jobs was involved with designing a circuit board for the Breakout game. You may remember the game; you bat a ball against a brick wall with a paddle and once you hit the wall enough times in the same place you can knock a hole through. Back then computer games were sold not as software, like today, but as a piece of hardware in a module that you plugged into the game console.

To reduce the cost of the modules Atari wanted a design with as few chips on as possible. Jobs subcontracted his work to Woz. Sound familiar? The master of efficient design had a project he could really sink his teeth into. Jobs had negotiated a bonus from Atari; $100 for each chip removed from the design. Woz astounded Atari by reducing the design by 50 chips earning Jobs and Woz $5,000. However, Woz didn't get his full share for reasons that tell a lot about both men.

Jobs told Woz that Atari had given him a bonus of $700 for the job that he split 50:50 with Woz giving him $350. Years later when the co-founder of Atari, Nolan Bushnell, published his autobiography Woz found out that his friend had got $5,000 from Atari and that he'd been short changed by $2,150. That's still lots of money today, and was a lot more in the 1970s. Woz was very philosophical about it saying in his autobiography that he didn't understand why Jobs had lied to him about the money, and though he was hurt he didn't see it as a good reason to end their friendship.

HOMEBREW AND THE BIRTH OF APPLE

Woz and Jobs were not the only people around with an interest in electronics and computers; after all we are talking about Silicon Valley. At this time a hobbyists club called the *Homebrew Computer Club* had started up, first meeting in someone's garage and then regularly at an auditorium of the *Stanford Linear Accelerator Centre* (yes, every neighborhood has a Linear Accelerator Centre, doesn't yours). Several founders of computer companies would be amongst its members, including Jobs and Woz.

Homebrew's mission was to bring computers to average people, to make it so people could afford to have their own computer. This of course was a perfect match with Woz and he became an enthusiastic participant. Everyone was really excited by a new microprocessor computer kit for sale called the Altair, from a small New Mexico company named *MITS*. You bought the kit for $379, put it together and you had your own computer.

Altair 8800 with 8-in. floppy disk system

Woz realized that the Altair was exactly like the little computer he'd designed and built 5 years before! The only difference was that the Altair had a single microprocessor, a CPU on one chip, and Woz's had a CPU that was on several chips. He also instantly realized that since chip technology had progressed he could now do much better and he started to design what become known as the Apple I. Really he just wanted to show off at Homebrew, just like he had at science fairs when he was a kid. Woz said, "*The theme of the club was 'Give to help others.' It was an expression of the hacker ethic that information should be free and all authority mistrusted. I designed the Apple I because I wanted to give it away for free to other people.*"

Before the Apple I, all computers, including the Altair, had hard-to-read front panels and no screens and no keyboards, you entered programs and data by laboriously flipping switches on the front, and you read the results from rows of red lights. After the Apple I, all computers would have keyboards and screens.

Sunday, June 29, 1975, was the big day. Woz had built and programmed everything. It was the first time that anyone had typed a character on a keyboard and seen it show up on their very own personal computer's screen. The Apple I was ready, but it had no name yet.

A fully assembled Apple I with a homemade wooden case

Ever the entrepreneur Jobs realized that people at Homebrew, and beyond, would want to buy Woz's circuit board. They would still have to solder all their chips to the board but they'd have a computer in a few days; moreover one that could use a small cheap TV screen and a keyboard. He encouraged Woz to form a company with him and suggested the name "*Apple Computer Co.,*" after the *All One Farm* commune he'd just returned from in Oregon. They both liked the name because it sounded fun and wasn't intimidating. Apple Computer had an interesting tension between the words, simple yet offbeat and it would get them near the front of the phone book.

Feeling that they might need some adult supervision the two Steves invited Ron Wayne, who drew up the partnership agreement, to join their new company. The Steves would get 45% each of the company and Ron the remaining 10%. The only real work Ron did for Apple was to design their first corporate logo, which depicted Isaac Newton sitting under an apple tree. In what has now become a legendary poor business decision, Ron sold his share back for $800 a couple of weeks later. He'd realized he'd be liable for any losses the new company racked up and, unlike the two Steves, he actually had some personal assets. It's now estimated that Ron's 10% of Apple would be worth over $20 billion!

Before Apple Computer could sell the Apple I, Woz insisted that he offered it to HP. Besides electrical engineering Woz senior had installed a very strong sense of ethics in his son and Woz had used equipment at HP labs, some components and probably quite a lot of HP's time in the design and build, so Woz felt compelled to offer his baby to HP.

The first Apple logo designed by Ron Wayne

They turned it down. What would HP do with a personal computer? To be absolutely sure Woz offered the design to HP legal and every other department, and they all turned it down. Jobs was free to start marketing the Apple I.

Woz and Jobs had shown the computer at Homebrew meetings several times, and people were impressed by it. Jobs found a Silicon Valley electronics storeowner who'd seen it at Homebrew and wanted to buy 100 Apple I computers fully built, for $500 each. Apple had its first sale for $50,000! Woz then had a difficult decision to make, he was now making more money from Apple than his full-time job at HP and so, very reluctantly, he quit HP. He'd dreamed of working there for the rest of his life, but now he could concentrate on designing his masterpiece—the Apple II.

The Apple II was going to be quite simply the best personal computer the world had ever seen. It would come in a sleek plastic case, the first time anyone had ever thought about making a computer look like a piece of chic consumer electronics. It would be expandable through expansion slots on the back allowing different hardware to be easily attached. Most importantly the Apple II was the first low-cost computer that you didn't have to be a geek to use.

An Apple II computer on display at the museum of the Moving Image in New York city

It was an instant huge success, partly because it was beautiful, partly because it was easy to expand and partly due to a program called *VisiCalc* that finally let a computer do something really useful. VisiCalc was the world's first spreadsheet application. It enabled business people to make financial models, engineers do complex calculations and everyday folks balance their household budgets. By 1980 Apple was the first company to sell a million computers and when they went to the stock market Apple were the biggest initial public offering since Ford. It made the most millionaires in a single day in history. The two Steves were rich beyond their wildest dreams.

Woz however, true to his father's principles, felt that there were many early employees at Apple who had not benefited from the company going public, though they had been crucial to its success. He sold a significant amount of his holdings to early employees at a big discount so they could share in Apple's fortune.

It's here we can return to Xerox PARC, because now Steve Jobs and a small group from Apple are being shown the Xerox Alto, with its mouse and revolutionary graphical user interface. Jobs could see that this was the future of computing and went back to Apple to design the Macintosh, a computer that would revolutionize computing.

Before we leave Apple for the time being, perhaps I should clear up some common untruths about Apple and Woz. He was not fired from Apple in some boardroom coup, or ousted by Jobs; in fact Woz is still an Apple employee and draws a modest salary from them, besides holding some very valuable Apple stock.

Woz had a private pilots license and was wealthy enough now to own his own plane. However, perhaps he wasn't a very good pilot because he had a bad crash during takeoff near Santa Cruz and suffered a very serious head injury. He had five weeks of total memory loss and was for a longer time unable to form any new memories. The medical term is: *anterograde amnesia*. So, not surprisingly he needed time off work to recuperate. He then went back to Berkeley to finally finish his degree, enrolling under the pseudonym Rocky Raccoon Clark. He also became involved in music promotion and bankrolled a huge outdoor music festival. It was the first to use big diamond vision screens and even had a satellite link to a sister concert in the USSR—it lost about $12 million. A second festival lost Woz even more money, but he could afford it.

Ever the pioneer Woz invented the first universal remote control, you know one remote that can replace all the remotes for the TV, video player, cable box and hi-fi. He started a company called "*CL 9*" to manufacture and market it, and it was bought out by another company after three years.

Woz also became a big sponsor of the arts and museums around Silicon Valley and in recognition the Mayor of San Jose named a street in his honor—*Woz Way* that leads to the *Children's' Discovery Museum* of which he was a major bene-factor. Children and learning have now become a major focus of Woz's life and he now regularly teaches fifth grade kids computing.[1]

A PC ON EVERY DESK

"I think there is a world market for maybe five computers." (Thomas Watson, Chairman of IBM, 1943)

You may be thinking by now that Apple was the only game in town. Of course that isn't true. The first computer that was ever offered to the public to buy was the *Honeywell H316 pedestal model.* It was marketed as the latest swanky kitchen appliance to be sold by the luxury US department store *Neiman Marcus*. The "*Kitchen Computer*" featured an integrated chopping board and was, "*useful for storing recipes.*" The perfect homemaker would have to take a 2-week course to learn to program the device, using only a toggle-switch input and binary light outputs.

---

[1] Woz has his own website at: http://www.woz.org

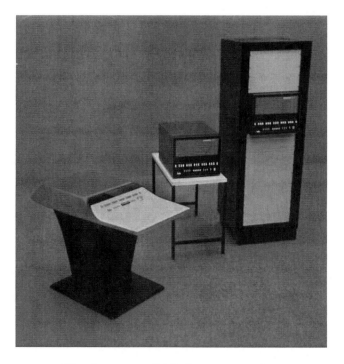

The honeywell kitchen computer

The full text of *Neiman Marcus'* advertisement read, *"If she can only cook as well as Honeywell can compute. Her soufflés are supreme, her meal planning a challenge? She's what the Honeywell people had in mind when they devised our Kitchen Computer. She'll learn to program it with a cross-reference to her favorite recipes by N-M's own Helen Corbitt. Then by simply pushing a few buttons obtain a complete menu organized around the entree. And if she pales at reckoning her lunch tabs, she can program it to balance the family checkbook. 84A $10,600.00 complete with 2 week program-ming course. 84B Fed with Corbitt data: the original Helen Corbitt cookbook with over 1,000 recipes."* Unsurprisingly, there is no evidence that anybody ever bought a Kitchen Computer.[2]

---

[2] A Kitchen Computer is on display at the Computer History Museum.

According to the *Computer History Museum*, the first PC was developed by John Blankenbaker of the *Kenbak Corporation* in 1970. The *Kenbak-1* didn't use a single microprocessor chip, but instead it had several small integrated circuits. It had just 256 bytes of memory and input and output was by a series of switches and lights. This computer was just like the first computer Woz had designed and built, and like his you couldn't really do anything useful with it at all. Costing $750 40 Kenbak-1s were sold; perhaps this was a lot to pay for a machine that really didn't do much.

The *Datapoint 2200*, also from 1970, was designed as programmable terminal for mainframe computers but when in 1972 it was upgraded to use the brand new Intel 8008 microchip it laid the foundation of the x86 architecture that would become the standard for PCs. Thus, the Datapoint 2200 is perhaps the best candidate for the first PC. It had a built in keyboard for input and a small screen for output and was about the same size as an electric typewriter. However, at $5,000 it was hardly affordable. To put this into perspective, a brand new Corvette sports car would set you back about the same money in 1970.

The Datapoint 2200

The development of PCs was not just limited to the US. The French claim that their *Micral-N* was the first commercial, non-kit set PC; it was also based on the Intel 8008 chip. However, the Micral-N sold for a staggering $13,000 about the same as a small house in 1970. Moreover, had you bought any of these PC's you wouldn't have recognized them as a computer since they had no user-friendly interface. As I described earlier, it was the Xerox Alto, developed at Xerox PARC in 1973, that was the first computer to have a mouse to move a pointer across a desktop that featured a GUI. If you had visited Xerox at this time, perhaps accompanying Jobs from Apple, you would have seen an interface that you would instantly recognize today with it's windows, icons, menus and folders.

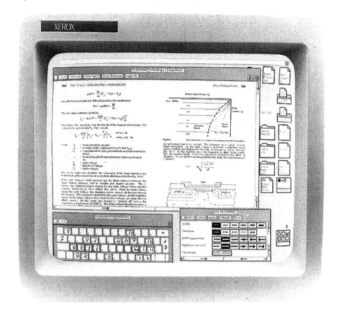

The Xerox graphical user interface

In many ways this interface was far ahead of its time and so well engineered for human-computer interaction. But, it would not enter ubiquitous use until the release of Windows 3.0 in 1990. Even now, in 2011, my MacBook is using basically the same interface. This is really a remarkable achievement. Think about it for a moment, 40 years later, although the hardware has changed dramatically, we are still using the same GUI. That was some invention. The Xerox Alto was never sold commercially. Its successor the Star was sold commercially, but was expensive and had little impact.

IBM around the same time were also developing a prototype PC called the SCAMP and in 1978 they released the IBM 5110, a larger descendant of the SCAMP. However, the 5110 would not be the breakthrough machine we'd all come to know as the PC. So as you can see by the mid 1970s numerous companies were all developing small computers intended for personal use, rather than use by a whole business or organization.

Three machines, the Commodore PET, the Apple II, and the Tandy TRS-80, all released in 1977, were referred to by *Byte* magazine as the "*1977 Trinity.*" In the late seventies, as silicon chip prices began to fall, many companies entered the home computer market: the Atari 400 & 800, the Commodore 64, the BBC Micro and the Texas Instruments TI-94 all sold millions and made profits until Commodore started an ill-advised price war.

Selling the Commodore 64 at cost was popular with consumers but by 1984 only Atari, Commodore and Apple were still left in the market. Moreover, since Atari and Commodore were strongly identified with the "*video game*" market (Atari's early success was largely due to the game PONG) there was little chance of a businessman ever buying these computers. This really left the business market to Apple who were quick to capitalize on their good fortune.

The IBM PC was released into the middle of this price war in August 1981. It had an open, architecture that allowed third parties to copy or *clone* it, and it used the new Intel 8088 chip. The XT added a huge (for 1983) 10 MB hard drive and could use up to 640 kB of RAM. IBM did *not* sell games and the target market for the PC was definitely businesses.

An IBM 5150

The majority of computers around this time would load their operating system from a floppy disk when they were turned on. The CP/M operating system was in widespread use along with BASIC as the standard programming language. In 1980, IBM approached *Digital Research*, for a version of CP/M to use as its operating system for the new PC. For various reasons the two parties were unable to agree to a contract. In, what is now considered one of the most fateful decisions in computer history, IBM approached Bill Gates, whose small company *Microsoft* was already providing BASIC for the PC. Gates offered to provide a disk operating system to IBM, which we will talk about in more detail later.

By the early 1980s the PC had gone from the plaything of hobbyists, like the Homebrew Computer Club, through mostly being used as video game consoles to a serious business tool that was rapidly becoming indispensable to any small business. In January 1983 *Time Magazine* named the home computer 1982s "*Person of the Year.*" It was the first time in the history of the magazine that an inanimate object was given this award.

KILLER APPS

PONG was the first *killer app*, but what is a killer app? In geek speak a killer app is a software application so popular that the desire to use it will make people go and buy the hardware on which to run the application (or app). Allan Alcorn was 24 years old when he designed the world's first popular video game. In 1972, just graduated and the second employee of new start-up Atari, he was tasked with designing a game to test out his programming skills. He invented an electronic table tennis game that was the precursor of all modern computer games. Alcorn later commented that:

> *"PONG was such a simple game that anyone could play. At that time, coin-operated games were dominated by pinball machines that had sometimes lurid graphics or driving machines that required skills that appealed to young males. PONG was unusual in that it required two players. I think it was the first game that appealed to young ladies and thus was a more social game."*

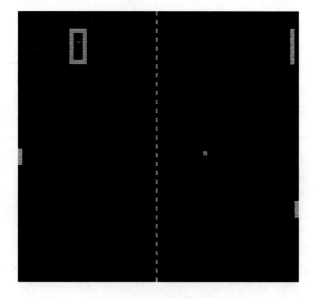

Atari's PONG game

The prototype arcade game PONG was installed in a small bar, called *Andy Capp's Tavern*, near Atari's office in Santa Clara, San Francisco, the area now known as Silicon Valley. The following day, a long line of people waited outside the bar for it to open, quarters at the ready. Pong was an instant hit. Reworked to run on an Atari home console PONG was the first time that you could use your own television set for anything other than just watching TV. PONG was a huge success and sales of the hardware needed to run PONG soared and by 1977 Atari's profit was $40 million. However, PONG was not going to encourage a businessperson to buy a computer. A businessperson needed a serious business reason for a PC. Spreadsheets and desktop publishing would provide this reason.

In the late 1970s Dan Bricklin, an MIT and Harvard Business School graduate, invented a visual calculating program based on the metaphor of an automatically updating table written on a blackboard, called *VisiCalc*. VisiCalc had all the same features as a modern spreadsheet, a grid of cells organized in rows and columns, automatic recalculation, status and formula lines, range copying with relative and absolute references, and formula creation by selecting referenced cells.

An example VisiCalc spreadsheet on an Apple II

Available initially for the Apple II, and then the Atari and Commodore, VisiCalc was the first business focused application for the home computer. *Lotus-1-2-3*, released in 1983 for the IBM PC then rapidly became the market leader, because of its greater power, speed and improved graphics. Executives who wanted to forecast sales and profits (and perhaps their own commission) finally saw an application they couldn't live without (a killer app). The IBM PC and Lotus-1-2-3 were now *must have* office tools. Lotus would continue their dominance of this market until Microsoft's own Excel spreadsheet eventually edged them out.

Ironically, although VisiCalc was first released for the Apple II, Apple let the IBM PCs and its clones eventually dominate the office financial application market. It would be a different story with desktop publishing, which we will come to later.

IT'S ALL ABOUT THE SOFTWARE

Until now we've mostly focused on the hardware, Babbage's Analytical Engine, the early computers like the Harvard Mark 1 and Colossus and the first PCs. Of course what makes these machines universal machines is as much about the programs they can run as the hardware they are made of. In a very real sense, as Turing proved, it is the program, the software, that makes them universal machines.

Perhaps the person who has been most identified with software since the PC started to boom in the 1980s is Bill Gates. Like so many of the other pioneers in this book his story is an interesting one.

Bill Gates, or William Gates III as he is really known, was born in Seattle, the son of a wealthy and prominent lawyer (William Gates II). From an early age he showed a keen interest in math and science. He probably first heard of computers when he visited the Seattle World's Fair in 1962 when he was six years old. The World's Fair theme was "*Century 21*" predicting what the world would look like and how people would live in the twenty-first century. At it's heart was the futuristic Space Needle connected to downtown Seattle by a monorail. If you've ever visited Seattle you'll have seen both of these, as they have become iconic landmarks.

Exhibits in the Fair's many pavilions predicted that we'd now all be driving flying cars, that we'd have personal rocket-packs and eat food from squeeze tubes like toothpaste. OK, so many predictions weren't quite right and we don't live like the *Jetsons*, but some predictions were spot on. People did land on the moon, we do have mobile cordless phones and we do use home computers for writing record keeping and shopping. Gates recalls that he loved the technical exhibits more than the funfair rides and that he visited every pavilion.

As we've seen so often with the early computer pioneers chance events conspired to give them opportunities that were not available to most kids. Gates was sent to the private Lakeside School in seventh grade. Unusually for 1968 Lakeside had a computer! Well, that's not strictly true; it had a teletype that was connected by a phone line to a PDP-10 minicomputer in downtown Seattle. I really can't over emphasize how unusual this was, not many schools had computer access in the 1960s.

Time on the minicomputer wasn't free, in fact it was expensive. However, the wealthy and well-connected *Lakeside Mothers Club* raised $300 to buy time on the computer. They thought this sum would last most of the year. Boy, were they wrong. Gates and his friends burned through the computer time in just a few weeks during which he wrote his first program to play tic-tac-toe, a lunar-lander game and even *Monopoly*; all in the computer language BASIC. Gates was very good at programming in BASIC. Whilst at Lakeside he became friends with an older pupil, Paul Allen, who also loved computing.

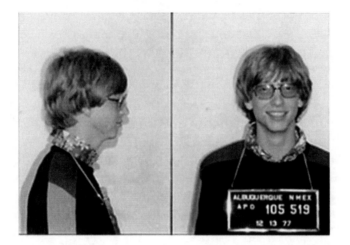

Bill Gates was photographed by the Albuquerque, New Mexico police in 1977 after a traffic violation

It wasn't long before Gates and his friends had become so skilled at programming the minicomputer that a local computer company paid them to track down and fix bugs in its programs. Well, they didn't actually pay the school kids but they gave them something just as valuable, free time on the computer. Gates was spending so much time on the computer working late into the night and the small hours of the morning that his schoolwork was suffering and his teachers and parents began to think he was becoming obsessed. He was forced for a while to stop using the computer and read books other than programming ones.

The fame of the Lakeside computer kids was spreading and a company in Portland Oregon employed them to write a payroll program in COBOL (remember the language developed by Grace Hopper). Again the boys were paid with free computer time but also royalties on the sales of the program.

Around this time Gates and Allen designed a program that could analyze data from traffic logging systems, called the *Traf-O-Data*. They formed a company and sold it to several towns making about $20,000 whilst still at school. They also got employed by a power company to debug their software.

Gates earned enough from this to buy his own speedboat. So it was whilst still at school that Gates realized he could make serious money from writing and selling software. Remember this was before most kids even knew what software was.

In the fall of 1973 Gates went to Harvard as a freshman, probably eventually to become a lawyer like his father. To be fair his parents didn't pressure him to become a lawyer, but they were very insistent that a young man needed a university degree if he were to succeed in life. Although Harvard offered many opportunities and distractions Gates gravitated towards Harvard's *Aiken Computer Center*, named after Howard Aiken, the designer of the Harvard Mark 1. The Computer Center had a PDP-10 minicomputer, the same computer he'd been using since he was a school kid.

His school friend Allen also moved out east to Boston when he got a job with the electronics company *Honeywell*. However, Allen's real motivation was to work with Gates on a new project that excited them both, not really to devote himself to his new employer. Gates and Allen had both seen the issue of *Popular Electronics* magazine that featured the Altair 8080 kit computer. Remember, this was the kit that so excited the members of the Homebrew Computer Club in Silicon Valley and inspired Woz to build the Apple I.

Gates and Allen were convinced that this was the start of the personal computer revolution and they wanted to be in at the beginning. Perhaps uniquely, they already knew that they could make money by selling software to people who bought computer hardware. The Altair had no software and they were convinced that they could get BASIC, which they were so familiar with from their school days, to run on the Altair. If they could, then everyone who bought the Altair would want a copy of their BASIC.

MITS based in Albuquerque New Mexico was manufacturing the Altair. So keen were they to get on the Altair bandwagon that Allen moved to Albuquerque to convince the owner of MITS, Ed Roberts, that he and Gates could write a version of BASIC for him. Gates, Allen and a new friend from Harvard, Steve Ballmer, formed a company to develop BASIC for the

Altair. They called the company *Micro-Soft* (yes with a hyphen).

MITS wasn't a big hi-tech outfit like Hewlett-Packard or IBM. It was a bit of shoestring operation run out of a tatty old building. They had made model rocket kits, which were very popular at the time because of the space-race and the moon landings, and had subsequently made a popular electronic calculator. Despite good sales, they'd never really made any money.

The Altair was to change that and in 1975 Roberts spent the summer touring the US in a blue camper van he called the *MITS-Mobile*. He visited hobby computer clubs, like Homebrew, demonstrating the Altair and Micro-Soft's BASIC. Sometimes Gates even went along.

PIRATES AND ENTREPRENEURS

Gates was now seriously considering dropping out of Harvard to concentrate his efforts on his new company and he had a new preoccupation, software piracy. Microsoft was one of the first companies to sell software to private individuals, people like the hobbyists at Homebrew; Gates became aware that many of them were swapping and copying BASIC for their kit computers.

This culture of sharing and borrowing and copying was at the heart of the hobby computer clubs and Gates famously published an open letter accusing them all of theft. The letter ended by asking people to send Microsoft money if they had copied BASIC for their computers. It seemed very simple to Gates, if they couldn't earn a profit from selling their software there would be no incentive for them to develop new and better software in future. This would ultimately be to the detriment of the hobbyists. Microsoft was a business not a hobby.

February 3, 1976

An Open Letter to Hobbyists

To me, the most critical thing in the hobby market right now is the lack of good software courses, books and software itself. Without good software and an owner who understands programming, a hobby computer is wasted. Will quality software be written for the hobby market?

Almost a year ago, Paul Allen and myself, expecting the hobby market to expand, hired Monte Davidoff and developed Altair BASIC. Though the initial work took only two months, the three of us have spent most of the last year documenting, improving and adding features to BASIC. Now we have 4K, 8K, EXTENDED, ROM and DISK BASIC. The value of the computer time we have used exceeds $40,000.

The feedback we have gotten from the hundreds of people who say they are using BASIC has all been positive. Two surprising things are apparent, however. 1) Most of these "users" never bought BASIC (less than 10% of all Altair owners have bought BASIC), and 2) The amount of royalties we have received from sales to hobbyists makes the time spent of Altair BASIC worth less than $2 an hour.

Why is this? As the majority of hobbyists must be aware, most of you steal your software. Hardware must be paid for, but software is something to share. Who cares if the people who worked on it get paid?

Is this fair? One thing you don't do by stealing software is get back at MITS for some problem you may have had. MITS doesn't make money selling software. The royalty paid to us, the manual, the tape and the overhead make it a break-even operation. One thing you do do is prevent good software from being written. Who can afford to do professional work for nothing? What hobbyist can put 3-man years into programming, finding all bugs, documenting his product and distribute for free? The fact is, no one besides us has invested a lot of money in hobby software. We have written 6800 BASIC, and are writing 8080 APL and 6800 APL, but there is very little incentive to make this software available to hobbyists. Most directly, the thing you do is theft.

What about the guys who re-sell Altair BASIC, aren't they making money on hobby software? Yes, but those who have been reported to us may lose in the end. They are the ones who give hobbyists a bad name, and should be kicked out of any club meeting they show up at.

I would appreciate letters from any one who wants to pay up, or has a suggestion or comment. Just write me at 1180 Alvarado SE, #114, Albuquerque, New Mexico, 87108. Nothing would please me more than being able to hire ten programmers and deluge the hobby market with good software.

Bill Gates

Bill Gates
General Partner, Micro-Soft

Bill Gate's open letter to hobbyists

The open letter was read out at a meeting of Homebrew and the members thought the notion was hilarious, though some were angered at being called thieves. Gates just didn't get it; software was for the common good and should be freely available to all, like songs on the radio were. The idea of software piracy was born here and still concerns the industry today.

Microsoft now set up an office in Albuquerque and started hiring more staff. Gates was an astute businessman; you don't get to become the world's richest man by accident. He realized that his company couldn't survive let alone grow if it was dependent on the Altair and hobbyists who didn't like paying for software. Gates persuaded large corporations like *General Electric* and *NCR* that they needed Microsoft BASIC to run on their existing computer systems.

Gates also sold BASIC for the new generation of personal computers like the Apple II, the Commodore and the Tandy. BASIC was soon the de facto standard programming language across the infant personal computer industry. Not content with that Microsoft diversified and produced a version of FORTRAN, the standard engineering and scientific programming language of the time.

By 1978 Gates decided that their original reason for being in Albuquerque, MITS and the Altair, was no longer the company's future and they needed to be closer to the action. So they moved back to their hometown Seattle, which was at least only a short flight from Silicon Valley.

THE DEAL OF THE CENTURY

I said, "*you don't get to be the world's richest man just by being lucky,*" but you do need to have some breaks along the way. IBM were about to hand Gates and Microsoft the opportunity of the century.

IBM were looking for an operating system for their new PC and they approached Microsoft. Now Microsoft didn't have an operating system at the time. Gates approached another software company, called *Seattle Computer Products*, who did have an operating system based on CP/M called *QDOS* for

Quick and Dirty Operating System. Gates bought the rights to QDOS for $50,000 with no strings attached. In what became one of the shrewdest business deals of the century Gates licensed it to IBM as PC-DOS, but Microsoft retained the right to license it to any other party as MS-DOS.

As copies of the IBM PC, known as *PC-clones* in the industry, flooded the world during the 1980s and 1990s the revenue from MS-DOS provided Microsoft with a huge income stream that enabled it to dominate the PC environment. Again, with hindsight, had IBM bought DOS from Gates outright, they could have denied the infant Microsoft the revenue stream from each DOS license. But you see IBM only expected to sell at most a few 100,000 PCs, they never expected tens of millions would be sold. Gates however had a vision of a computer sitting on every desk and of Microsoft's software being on each of those computers.

Gates had his software, DOS and BASIC, running on every IBM PC and PC clone in the world; he also realized that the killer app for these computers was the spreadsheet. He was not about to leave this lucrative market to VisiCalc and Apple and so in 1982 Microsoft produced their own spreadsheet called *Multiplan* that would eventually evolve into the market leading *Excel*, which I'm sure many of you are familiar with.

You're probably wondering why the developers of VisiCalc didn't have a patent on the spreadsheet and stop Microsoft copying their idea. Well, unfortunately back then it was not possible to patent software. The patent process did not apply to something that was held to be just an idea, there was no physical reality to software, it would be like patenting a piece of music. You could copyright software, but not patent the idea behind it. Thus, Microsoft could not directly copy VisiCalc and sell it as Multiplan but they could program their own spreadsheet and sell it—the new program would not legally be a copy since its code would be different.

It seems here that Gates' ethics are questionable. Hobbyists copying his software infuriated him, but he wasn't concerned about profiting from somebody else's ideas. This ethical issue is at the heart of today's open source software movement, and proponents of open source software would argue that Microsoft should have worked with VisiCalc to improve their original product for the benefit of everyone.

Having created a spreadsheet, the next essential piece of business software was of course the word processor and Microsoft produced *Word* in 1983. Everything was going well for Gates, he was even selling Word to Apple II users. Gates was settling down in his luxury lakeshore house to watch the Super Bowl on January 22 1984 when everything changed.

1984

We left the two Steves of Apple after Jobs had visited Xerox PARC and seen the GUI of the Altos. Jobs realized that the GUI was the future of computing and went off to design a new computer around this revolutionary way of interacting with computers. Apple's new product would be the Macintosh, named after a variety of apple, and it was now ready to launch to the public.

TV advertising spots during the Super Bowl are the most coveted, and expensive spots advertisers can buy. Apple decided to take a revolutionary approach to advertising their new product. Instead of running a traditional campaign, with many TV commercials spread over weeks or months they would run a single high quality commercial, almost like a movie trailer, just once in a break in play during the Super Bowl. The commercial would never be shown again.

Ridley Scott, who had recently directed the hit science-fiction movies *Alien* and *Blade Runner,* directed the commercial. The totalitarian world depicted in the commercial is of course a reference to George Orwell's famous book *1984*, about a dictatorship where "*Big Brother,*" the face on the screen, tells the workers what, and how, to think and controls every aspect of their lives. Big Brother is a reference to IBM, sometimes called "*Big Blue,*" and the grey workers are people who have been forced to buy and use IBM PCs.

Apple's message is clear, think for yourself, free yourself, and use a Macintosh. Of course nobody knew what a Macintosh was on the day of the Super Bowl. Apple's bold advertising decision worked. The following day everyone was taking about the commercial and the Macintosh. In fact, the "*1984*" commercial, as it is called, is still widely regarded as the one of the best TV commercials ever. In 1999 *TV Guide* rated it the "*Number One Greatest Commercial of All Time*" and in 2007 it became the "*Best Super Bowl Spot,*" in the games' 40-year history. Of course all this great publicity would have been wasted if the Macintosh didn't live up to the hype. Fortunately for Apple, eventually it did.

An original 1984 Apple Macintosh

Everything about the Mac was different to the computers that preceded it. It was small and friendly. The screen and computer were one unit, not like the large heavy separate boxes and monitors of the IBM PC's. There was a mouse and most importantly the revolutionary GUI. For the first time a person could buy a computer, take it home, take it out of the box, turn it on and use it without having to learn and type complex and arcane commands. If you saw the interface on the original Mac today you'd quickly be able to figure how to use it as its GUI is the ancestor of all of the interfaces you may have used: Window, OS X or Linux.

I have a personal anecdote about how unfriendly IBM PCs were. Back in 1989 I started my first research job and took delivery of a new PC. I unboxed it, connected all the cables and switched it on, but nothing happened. Like most men, when all else fails I read the instructions. It told me I had to first install DOS from the installation floppies it shipped with. Can you believe computers were sold without the operating system factory installed! So, I started the installation process by putting the first DOS disk in the floppy drive and turning the computer on. It crunched away for a while and then asked me to *"Remove the installation disk"* and then *"Insert a blank formatted disk."*

Now this was a problem. I didn't have a blank formatted floppy disk, the computer hadn't shipped with one, and to format a floppy disk I'd need to have DOS installed. I couldn't install DOS *without* a formatted floppy—Catch 22! Fortunately, I was working at a university, and it was easy to pop down the corridor and borrow a blank formatted floppy to continue the DOS installation process. Really, what were the people who designed this process thinking? If you'd bought a PC and taken it away off to your refuge in the wilds you'd be buggered when it asked you for that formatted floppy.

Not only was the Mac easy to use it, it also featured the next killer app—desktop publishing. Mac users could create complex page layouts, using a program called *MacPublisher,* and see their pages directly on screen just as they would look when they printed. This is called "WYSIWYG," for What You See Is What You Get. The ability to create WYSIWYG page layouts on screen and then print pages at a crisp 300 dots per inch resolution, with Apple's new laser printer (another idea borrowed from Xerox PARC), was revolutionary for the publishing industry, providing a flexible and lower cost alternate to commercial phototypesetting. In 1985 *Aldus PageMaker* was released and it quickly became the publishing industries standard software.

The Apple Mac became a *"must have"* if you were serious about publishing. For the first time you could create complex page layouts with columns, sidebars, box-outs, and pictures, experiment with your design as much as you liked, then hit *"Print"* and almost instantly hold your publication in your hands. Anyone with a Mac and a laser printer was now a publisher.

The Mac's revolutionary interface, based on the GUI that Steve Jobs had seen at Xerox, was essential to the success of desktop publishing. It just wasn't possible to create and edit a complex page layout without a GUI. Since the IBM PC and clones lacked a GUI, desktop publishing software wasn't available for the PCs.

It's a good story, the one of Bill Gates watching the Super Bowl in January 1984 and being totally surprised by Macintosh, but of course like all these legends its not strictly true. You see, Microsoft had been working with Apple developing software to run on the Macintosh for some time, so they were quite familiar with its GUI. Gates was however quick to realize that Microsoft needed a GUI of its own if it was to compete with Apple. However, developing a useful GUI would be much harder than he thought and it wasn't until the release of Windows 3.0 in 1991 that Microsoft and the PCs had a workable GUI. By then though, Macs were firmly placed on the desktops of everyone in the publishing and *"creative"* industries and these industries have never really crossed back over to the PC platform.

"To create a new standard, it takes something that's not just a little bit different; it takes something that's really new and really captures people's imagination—and the Macintosh, of all the machines I've ever seen, is the only one that meets that standard." —

Bill Gates

After PONG, the spreadsheet and desktop publishing we'd have to wait until the invention of the Web before another killer app was born, making a whole new generation of users go out and buy a universal machine.

Chapter 8

WEAVING THE WEB

*I*n the summer of 1993 I was working with a small team developing expert systems (artificial intelligence software) for the construction industry when Mark came into my office. He was waving a floppy disk and enthusiastically said "*Ian, you've got to see this! It's a web browser!*" "*What's the Web and why do I want to browse it?*" I replied. "*It's really cool, you can see information from all over the world and navigate around it like a web,*" he said, so I took his disk and installed a browser called *Cello*, and then had to install some other network drivers, and after about half a day of lost work I was ready to browse the Web.

I launched Cello and I was taken to *http://info.cern.ch*, which seemed to be the heart of the Web. I then browsed around CERN getting lots of arcane documents about particle physics experiments and committee meetings, and ended up in a Computer Science department at MIT. I browsed around other websites for an hour or so and then put Cello down. Frankly, the Web seemed rather boring.

Less than a year later Mark was back, waving another floppy saying, "*Ian you've got to see this, it's a new web browser called Mosaic!*" I told him "*I'd browsed the Web and found it dull.*" "*No! Now it's got pictures!*" So I installed Mosaic and browsed the Web again. It was remarkable how much the Web had grown and changed in a few months. Now many students, mostly in computer science departments, had their own web pages, and there were photos of things like their dog or their mountain bike, and there were links to shops that stocked their favorite stuff like snow boards and tech books. Everything was very rudimentary and basic; mostly just black text on white backgrounds, a few grainy

I. Watson, *The Universal Machine*,
DOI 10.1007/978-3-642-28102-0_8,
© Springer-Verlag Berlin Heidelberg 2012

photos and lists of links to click on. I could see, though, that this was very personal information and could become really useful if it were to ever break out of academia. How the Web broke out is the subject of this chapter.

VAGUE BUT EXCITING

In March 1989 a young scientist, Tim Berners-Lee, working at the European atom-smashing lab CERN, wrote a research proposal that showed how information could be accessed and easily shared over the Internet by using *hypertext*. This is a way of navigating or browsing through information by clicking on links. Every web page is an example of hypertext, and I'm sure you are very familiar with browsing by clicking on links in webpages. In the 1980s research into hypermedia was all the rage; there was even a very popular software application for the Macintosh called *HyperCard* that allowed regular people (that is non-programmers) to easily create small hypermedia applications for the Mac.

Berners-Lee's big insight was to see that the Internet could be used to link hypertexts together. For example, a webpage at CERN in Geneva could link to a webpage at MIT in Boston, which in turn could link to a page at Caltech in Pasadena. The physical location of webpages would be irrelevant to users, as the Internet would provide a seamless transport system for the pages. Another key innovation was that users didn't need any form of permission to link to a webpage – all they needed was its address. If you could view a webpage then you could link to it. In this way individual users all over the world could curate their own small document collections that could be linked together into a global information resource of massive and unprecedented scale. Berners-Lee's boss at CERN, Mike Sendall, wrote "*vague, but exciting*" on the research proposal and it was allowed to proceed.[1]

---

[1] Brian Carpenter told me that Mike Sedall and his boss David Williams said, "*our important contribution to the development of the Web was not getting in Berners-Lee's way!*"

Sir Tim Berners-Lee in 2010

I'm fortunate that a colleague working in my department, Brian Carpenter, worked with Berners-Lee at CERN and he recalls that:

"In 1984, Tim Berners-Lee, who had so impressed Robert Cailliau and me during his spell as a contract programmer in 1980, applied to CERN for a Fellowship, the same route by which both Robert and I entered the lab. He was hired by the On-line Computing group, whose job was to provide soft-ware and support for the data acquisition systems run by every physics experiment at CERN...as usual with Tim, his day job wasn't enough, and he looked at another major problem facing CERN: how on earth could we manage the mass of disparate information and documentation needed by CERN itself and by the enormous experiments? Without exaggeration, thousands of people were writing documents, and anyone in the physics community might need to find any of them at anytime.

This wasn't a new problem; it was generally felt in the networking world that some sort of information infrastructure was needed, and there were several solutions appearing around the Internet. But Tim, working with Robert, put together the ideas of hyperlinks, standardized document markup, and remote calls over the network – the first being the idea he had tested out in his old Enquire program, and the second two being ideas that Robert and I had introduced him to in 1980. His original proposal for the web landed in my mailbox in March 1989; I scribbled a few comments on it and sent it back. I didn't hear much more until one day in early 1991 when Tim popped into my office and said 'Are you logged in to PRIAM? [our main Unix host at the time]... Good, type WWW.' So I did, and up came the world's first simple web browser – not a screen full, just a single line of text listing a few topics. Tim had already arranged for several very useful databases to be available to the web, such as the CERN internal telephone directory. I used it from then on to look up phone numbers – it was quicker and more accurate than the printed phone book.

A bit more than two years later, in the summer of 1993, Tim popped into my office again, with the same opening question. By then, instead of a plain old terminal, I had an X-Windows workstation, which provided access to Unix with proper graphical windows, using Internet protocols. Tim got me to call up a window showing the just-released Mosaic graphical web browser. I've never had a day at work without a browser window open since then. The Internet had finally found its killer app, and its universal growth instantly became inevitable." – Brian Carpenter

Berners-Lee and Robert Cailliau built the first web server and browser in May 1990 and came up with the impressive name, the WorldWideWeb (WWW for short). *http://info.cern.ch* was the web address (URL) of the first webpage running on a NeXT computer at CERN.

World Wide Web

The WorldWideWeb (W3) is a wide-area hypermedia information retrieval initiative aiming to give universal access to a large universe of documents.

Everything there is online about W3 is linked directly or indirectly to this document, including an executive summary of the project, Mailing lists , Policy , November's W3 news , Frequently Asked Questions .

What's out there?
 Pointers to the world's online information. subjects , W3 servers, etc.
Help
 on the browser you are using
Software Products
 A list of W3 project components and their current state. (e.g. Line Mode ,X11 Viola , NeXTStep , Servers , Tools ,Mail robot ,Library)
Technical
 Details of protocols, formats, program internals etc
Bibliography
 Paper documentation on W3 and references.
People
 A list of some people involved in the project.
History
 A summary of the history of the project.
How can I help ?
 If you would like to support the web..
Getting code
 Getting the code by anonymous FTP , etc.

A screen shot of the first web page around Nov 1992[2]

Berners-Lee didn't invent hypertext; it can be traced back to Vannevar Bush and his *memex* machine. In the *Atlantic Monthly* in 1945 he wrote: "*Wholly new forms of encyclopedias will appear, ready made with a mesh of associative trails running through them, ready to be dropped into the memex and there amplified.*" Of course, as we've seen, Vannevar Bush had no idea how to build a memex back then. Doug Engelbart, inspired by the memex, developed, possibly the first, hypertext system at SRI in the 1960s and demonstrated it during "*The Mother of All Demos,*" which I described in chapter six. If you would like more information on how the Web was developed and its impact Tim Berners-

---

[2] http://www.w3.org/History/19921103-hypertext/hypertext/WWW/TheProject.html

Lee has written an excellent book, "*Weaving the Web,*" whose title I borrowed for this chapter.

EXPLOSIVE GROWTH

In 1993 an MIT researcher counted 623 websites[3]; some of these were the ones I had browsed with Cello. By the end of 1994 there were more than 10,000 websites. These no longer just carried serious academic stuff; there was the totally pointless "*Amazing Fishcam*" that took photos of a fish tank every few minutes. *Fogcam* at San Francisco State University mostly showed a grey wall of fog or, on a good day, part of the campus. It's now the world's longest running webcam. You could order pizza online from *Pizza Hut* in Santa Cruz; *First Virtual* was the world's first cyber bank, and early signs of the unorthodox nature of the Web were beginning to show with sites like the *Museum Of Bad Art* and *JenniCam.*

JenniCam was started in 1996 by a young student, Jennifer Ringley, who put a webcam in her dorm room that filmed her 24/7; including sometimes naked and having sex. Claimed by some to be conceptual art it was one of the first *lifecasting* sites and was a precursor of reality TV shows like *Big Brother.* JenniCam claimed 100 million page views a week and for a while it made her a celebrity. It was quickly joined by websites with a more transparent purpose like *Bianca's Smut Shack*, and *Sex.com.* It was clear that the Web would deliver information on *anything* that interested people.

In the mid-1990s I watched with interest, from my office in England, as the Web grew. This growth was mostly in the US, centered on the east and west coasts around the great university campuses. The Web didn't really seem to have taken root in England yet. At this time I was consulting for a US company based in the San Francisco Bay Area. They used to fly me out a couple of times a year, which I enjoyed.

---

[3] http://en.wikipedia.org/wiki/List_of_websites_founded_before_1995

I remember in about 1997 walking in downtown San Francisco when I saw a huge billboard on the side of a building. I can't remember what it was advertising, but I was amazed to see below the photos and captions a URL: *www. companyname.com* (as I said, I can't remember the company name). I remember stopping and thinking *"Wow! Companies have websites here and use them for advertising, this is going to be huge!"* I wasn't wrong, but sadly I didn't quit my job and start up a dotcom company that would make me a billionaire overnight. Some people did though, but before we can tell their story we must find out more about the Internet.

A NETWORK OF NETWORKS

The US military once again has a major role to play in the early development of what was to become the Internet. When the Soviet Union launched *Sputnik,* the first satellite, in October 1957, it shocked the world and particularly the US. The US thought that it was the most technologically advanced power on Earth; how could the Soviets have beaten them into space? The US responded by creating the *Advanced Projects Research Agency* (ARPA) to funnel research money into regaining its technological lead. This was the height of the Cold War and the US military was concerned about the surviv-ability of its communications networks during a Russian nuclear attack. ARPA funded a project to network the nation's computers together in what would eventually become known as the *ARPANET.* The technological innovations developed for it became the basis of today's Internet.

There were two important innovations in the ARPANET: *hier-archical routing* and *packet switching*, which I'll describe in turn. The US military wanted to be able to connect all its new command and control computers together in a way that could survive a Soviet nuclear attack. Well, so the popular origin story of the ARPANET goes, but like many a good story in the history of computing, it's a myth. In reality the ARPANET was designed to connect scarce computing resources together so that, for example, a scientist in Caltech, California could use a powerful computer at MIT in Boston. They wanted a network that was reliable and extremely fault tolerant; it wouldn't all stop if one part were broken. Now it so happens that this was also what the military wanted, and so the myth that the military commissioned the ARPANET to survive a nuclear attack was born.

If you're designing a computer network, ideally you want every computer to be able to communicate with every other. So let's build a network where every computer is physically linked by a cable to every other computer.

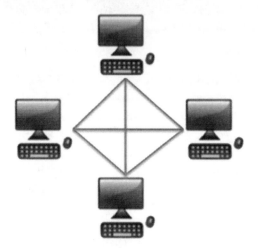

Every computer linked to every other computer

I think you can see that this network is very efficient, because you can send data to any computer in the network in a single *hop*. However, although efficient in transmitting data, it is very inefficient in the amount of network cable used. Imagine if you wanted to connect thousands of computers this way, you'd need an awful lot of cable. If we wanted to minimize the amount of cabling, a simpler arrangement is where each computer is only connected to its neighbor in a chain.

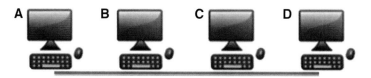

Computers networked in a chain

This configuration uses very little cable but has two dis-advantages. First, if computer "*A*" wants to send data to "*D*" it will take three hops. Second, if computer "*B*" malfunctions, perhaps because a Russian bomb has taken it out, the network is broken and "*A*" is isolated from "*C*" and "*D*". There is no *redundancy* in the network. A more redundant configuration might be a ring.

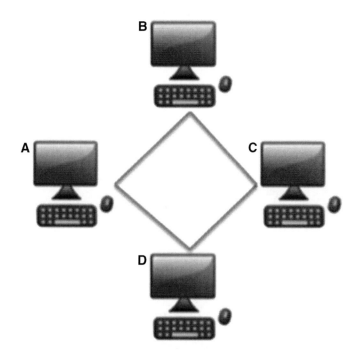

Computers networked in a ring

Here if computer "*A*" wants to send data to computer "*C*" it can do so in two hops via computer "*B*" or "*D*." Now if computer "*B*" is down, "*A*" can still communicate with "*C*" via "*D*." We've added some cable we didn't strictly need but gained some redundancy. The ring network is still vulnerable if it is broken in several places; moreover if the ring is very large we might have a lot of hops to get data around it. Finally, we could use a hierarchy of networks, or a network of networks.

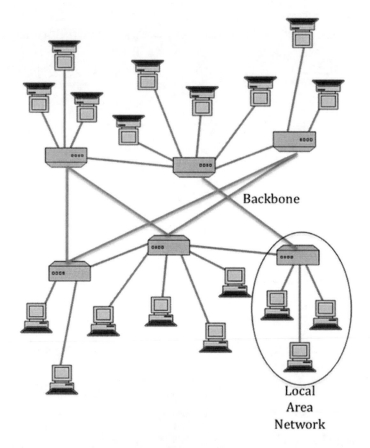

Backbone

Local
Area
Network

A hierarchical network

Here there are a series of *local area networks* (LANs) that are connected to other LANs in their vicinity. The set of LANs are in turn connected to the *backbone*. The backbone can carry much more data than the local area cables and can have some redundancy built in to it. Sure, a LAN might get disconnected if a Russian bomb landed on it, but the rest of the network would be unaffected. This basically is the architecture that was worked out for the ARPANET and is the architecture of today's Internet. Your house may contain a LAN if you have several computers sharing the same Internet connection, perhaps via Wi-Fi. Your broadband router or modem is connected to your *Internet Service Provider* (ISP), the next level up in the hierarchy, and your ISP is connected to the backbone, usually by fiber-optic cable capable of transmitting a vast amount of data at very high speeds.

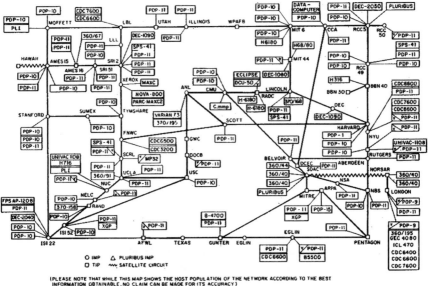

Logical map of the ARPANET, March 1977

The second key breakthrough was *packet switching,* invented independently by Donald Davies in the UK and Paul Baran in the US. When you call someone on a landline telephone, a complete electrical circuit is established between

171

your handset and that of the person you are calling.[4] Back in the past you could see telephone company operators manually creating these circuits by plugging electrical cords into sockets. This is a very wasteful use of a network because when you are speaking there are frequent pauses, when nothing is being said and no information is being transmitted, but the line is totally reserved for your conversation. Even though the telephone company can route numerous calls simultaneously over a single phone line, through clever electronics, it's still a very finite resource.

In packet switching no single communication has total control or use of a single path through the network; there is no dedicated user-to-user connection. Imagine you have some data to send across a network like an email. The email has an address, the recipient: *j.doe@company.com,* and the body or text of the mail. This data is not sent as you might think in one piece, but rather it is divided into *packets,* each of which is sent separately. Let's use an analogy; think of data like a loaf of sliced bread, each slice of bread is a packet that is numbered "*1 of 30*", "*2 of 30*", etc. The slices are sent across the network to the recipient individually. Each slice may take an entirely different route and may travel at different speeds, arriving at their destination out of order. At the destination the recipient's software puts the slices back in order. If a slice is missing, got lost, misrouted or got held up in a virtual traffic-jam the recipient can ask the sender to: "*resend packet #12*". When the recipient has got all the packets in the correct order, the data (our email) is complete and is successfully delivered. Packet switching makes much more efficient use of the network because the packets can be routed around faults so everyone can use the network simultaneously and much closer to its maximum data-carrying capacity.

In addition to the cabling and packet switching the ARPANET team, led by Bob Taylor, had to build the hardware and software that would enable computers, called *nodes*, on the network to

---

[4] Modern telephone networks are now switching to use packet switching and VOIP (Voice Over Internet Protocol).

communicate with each other. The hardware that made the network possible were small computers called *Interface Message Processors* (IMPs). Well, they were small by the standards of 1968, but they seem huge today. The IMPs took care of the packet switching and forwarding, and today have been replaced by *routers*. They were connected to modems and leased telephone lines. Individual computers were connected to the IMPs.

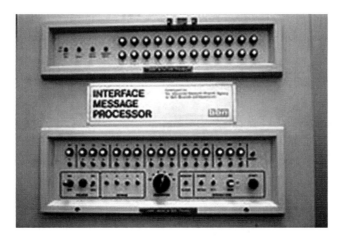

Front panel of the first IMP

The first message on the ARPANET, "*lo*", was sent at 10:30 pm on October 29 1969 by a UCLA student called Charley Kline. He had been told to connect to a computer at SRI. The message was supposed to be "*login*," but the system crashed after just "*l*" and "*o*" had been transmitted. An hour later, after the computers had been rebooted, "*login*" was successfully transmitted and received. By December 1969 there were four nodes on the network, all on the west coast. The east coast was connected the following year and by 1973 there were 40 nodes, including university, government and military computers. Norway, the first country outside the US to be connected, joined in 1973 soon followed by England. The ARPANET continued to grow until 1983, when the US military nodes split off into a private military network called MILNET.

In 1991 a democratic Senator, Al Gore, drafted the *High Performance Computing and Communications Act*, often called *"the Gore Bill"*, which was passed into law, helping the creation of the modern Internet, or the *"information super-highway"* as Gore called it. This has caused some people to state, sometimes in jest and sometimes seriously, that *"Al Gore invented the Internet!"*

YOU'VE GOT MAIL!

The Internet isn't an application, it's infrastructure like roads, sewers and railways. Just as cars, buses and trucks can run on the same roads, different applications run on the Internet. Seventy-five percent of the traffic on the ARPANET was email, and email remains one of the primary uses for the Internet today. Email predates the ARPANET, since users on time-sharing mainframes were able to communicate with each other by sending messages to other users on the main-frame. There might be hundreds of users with accounts on a single mainframe, or *host*, and you could send a mail to any user's account. This was a convenient way of sending messages to other people within, for example, the head office of a bank or a university, but you could not send mail to users on a different host.

This was clearly a limitation as large organizations, like universities, may have several computers; people also may want to communicate with colleagues in other organizations. Researchers at the City University of New York and Yale devel-oped a point-to-point network to link their computers together. Called *BITNET*, for *"Because It's There Network"* other universities were encouraged to join. All you needed was a leased telephone line connected to a BITNET node and a modem. In Europe IBM encouraged BITNET's use to link its computers and it was called *EARN*; in Canada it was called *NetNorth*, in India *TIFR* and *GulfNet* in the Middle East. Even-tually thousands of computers were attached.

Other computer manufacturers developed their own similar networking solutions to allow their computers to communicate with each other, but the different networks didn't communicate with each other. These information islands were what the ARPANET was designed to break down. Providing your computer could communicate with an IMP, the IMPs would communicate with each other. The stage was set for communication between computers regardless of their manufacturer.

Ray Tomlinson is credited with sending the first email and with deciding to use the now familiar "@" symbol. He says that using "@" for "*at*" was entirely logical, since to address a mail you needed the recipient's account name and the name of their computer or host on the network. The IMPs would ensure that the email was delivered to the correct host, and software on the host would deliver the mail to the correct user account. So, *user_accountname@hostname* is still what we use today for email addresses. Tomlinson sent the first email between two DEC-10 computers that were actually located right next to each other, so the mail didn't really have far to travel. Tomlinson says he can't remember what the content of that first email was; probably something random like "*qwertyuiop.*" In 2010 it was estimated that 107 trillion emails were sent.[5] Tomlinson never made any money from his invention, but imagine if he got a royalty of one cent on each email. That would be a nice little earner!

AN ELECTRONIC NOTICE BOARD

The ARPANET and email were very useful if you worked for a university or large corporation, but of course most people didn't and so had no access to the network. As personal computers started to become popular in the late 1970s and 1980s, home users increasingly wanted to contact other users and share information. In 1978 Ward Christensen, apparently whilst snowed in during a blizzard in Chicago, came up with an idea for a computerized bulletin board system. Like a physical notice board, or *bulletin board*, users would be able to read

---

[5] http://royal.pingdom.com/2011/01/12/internet-2010-in-numbers

notices posted on the bulletin board, and users with the correct privileges would be able to post notices and edit them. A user would login using their home computer, which would dial up the bulletin board using a modem connected to a standard telephone line.

Back then modems were very slow, but that didn't really matter because the bulletins only contained plain text. Another drawback was that since long distance calls cost money, to be affordable the bulletin board system (BBS) had to be local to its users. Thus, BBSes were set up all over the US and overseas.

Welcome screen of Neon#2 BBS (Tornado)

Users of a BBS could send email to other members, chat with them online, play simple computer games, upload and download files, and participate in specialist forums. Most BBSes were run as a hobby by geeks and mostly dealt with computer-related issues, but some were more general in scope. In 1966, Stewart Brand, a leading member of San Francisco's counter culture, had published an alternative lifestyle catalogue called the "*Whole Earth Catalog.*" This was the bible of the hippie commune movement around the Bay Area, and updates were printed sporadically into the 1970s. In 1985 Brand started a BBS called *The Whole Earth 'Lectronic Link*, or The WELL for short. It was a place for members to share information about all manner of subjects ranging from restaurant reviews in San Francisco to the music of the Grateful Dead. As we will see in a later chapter, BBSes also became an essential virtual meeting place for hackers.

A major limitation of BBSes is that they were isolated from one another. Although some attempts were made to let them communicate with each other, this was always on a fairly ad hoc basis. However, large online communities did build up using systems such as *America Online* (AOL), *Prodigy* and *CompuServe*, which had 600,000 members in 1990. However, during the 1990s BBSes slowly lost their popularity as the Internet and the Web took over. BBSes either had to change into web-based systems, which AOL successfully did, or face extinction. A web version of The WELL still exists.[6]

THE BROWSER WARS

In 1995 Microsoft released *Windows 95*, in one of the largest software launches ever seen. A $300 million advertising campaign included the Rolling Stones, and Jennifer Aniston and Matthew Perry, from the hit TV comedy *Friends*. The Empire State building in New York was lit up in the colors of the Windows logo and there were synchronized launch events around the world. If you bought the *Plus!* add-on pack for Windows 95 it included Microsoft's first web browser, *Internet Explorer 1.0*.

Microsoft hadn't actually built its own browser, but had licensed the code from Mosaic and reworked it to make it look like a Microsoft product. Marc Andreessen and Eric Bina had developed Mosaic at the *National Center for Super-computing Applications* at the University of Illinois Urbana-Champaign. Versions of Mosaic ran on Unix, the operating system many mainframes and minicomputers used, as well as popular home computers like the Commodore Amiga, Macintosh and Windows PCs. In 1994 *Wired* magazine wrote: "*The Revolution Has Begun: Don't look now. But Prodigy, AOL and CompuServe are all suddenly obsolete – and Mosaic is well on its way to becoming the world's standard interface.*"

---

[6] http://www.well.com/

Despite having hundreds of thousands of members, the old BBSes were obsolete because of the Web's fundamental advantages. These were first that the Web was free; once you were online you weren't charged for content, unlike for example AOL, which charged a monthly subscription fee. Secondly, on the Web you could access all the content that was there, whereas on AOL you could only see AOL's content, not what was curated by CompuServe. Thirdly, it was very easy to create your own website; if you could use a word processor then you could create a basic website. Moreover, nobody would stand over you and censor your content, or tell you what you could or could not put on your website.

Mosaic had become the killer app for the 1990s. In 1993 there were about 600 websites; by the beginning of 1997 it is estimated there were 650,000. The Web was receiving almost unparalleled attention in the media, and gurus and pundits were talking of a "*second industrial revolution*" and of a "*new information revolution*" whose impact would be greater than that of the printing press. Understandably Microsoft, which totally dominated the PC market, wanted its piece of the action, which is why it licensed Mosaic for Internet Explorer and then quickly began improving on it.

Andreessen left academia and started *Netscape Communications,* producing Netscape Navigator in 1994. Netscape, like the web, was free at least for educational and non-profit institutions – businesses would have to pay. Netscape had one big innovation; it could display web pages as they were downloading, meaning users didn't have to look at a blank screen while a large webpage slowly downloaded. Remember, in the 1990s modems were still very slow.

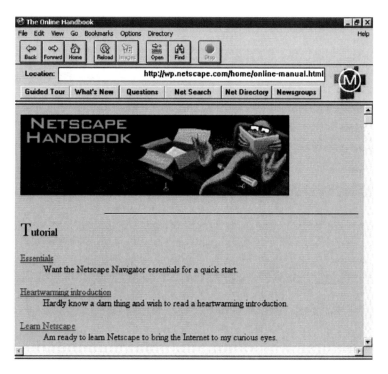

Netscape navigator V0.9

By the start of 1996 Netscape had almost 80% of the browser market[7] and Microsoft began to fight back. Internet Explorer (IE) 2.0 was now free like Netscape. Microsoft started doing deals with PC manufacturers to bundle IE along with Windows on new PCs, and finally Microsoft included IE with every copy of Windows, which had over 90% of the PC operating system market. Microsoft's logic was quite simple; most members of the public wouldn't bother to download Netscape if they already had a perfectly good web browser to use. Microsoft would therefore dominate the Web, just as it did the PC software market.

Why is the make of browser so important? IE, Firefox, Chrome, Safari, Opera all look much the same and all let you

[7] http://en.wikipedia.org/wiki/File:Netscape-navigator-usage-data.svg

browse just fine – it's about eyeballs. When you first open up IE it automatically takes you to the *Microsoft Network* (MSN) homepage and encourages you to login to your MSN account or create one if you don't already have one. Chrome takes you to Google when you first open it, and Safari to Apple. Now of course you can change your browser's *homepage*, the page it first displays when it opens, to any page on the web you like, the *New York Times*, or *Manchester United's* webpage. But you see, many users don't bother to change the home page, they just keep the default one. This means that if a significant proportion of users keep the default home page of IE, then Microsoft can sell other services, products and advertising to them. Put simply, the bigger your share of the browser market, the bigger your potential revenue.

The browser wars led to the *United States vs. Microsoft* antitrust legal action in 1998 that after two years of litigation found that Microsoft had used its monopoly of the PC operating system market to unfairly dominate the browser market and eliminate its competition. Paul Maritz, Microsoft Vice President, was quoted during the trial of wanting to, "*extinguish and smother*" Netscape and wanting to, "*cut off Netscape's air supply*" by giving away free clones of Netscape's product. The judge's ruling was that Microsoft should be broken up into separate parts, one selling operating systems and the other making software applications. Needless to say, Microsoft immediately appealed. The appeal process dragged on for a further four years, and by the end of all the legal wrangling it wasn't really clear what penalty Microsoft incurred. Netscape however was dead in the water; IE had nearly 95% browser market share by 2003. But Microsoft's dominance of the Web would not last for long and it would soon be challenged.

A NEEDLE IN A MILLION HAYSTACKS

As the number of websites grew explosively, finding pages on a topic you were interested in became more and more challenging. In the early days of the web people maintained and published lists of their favorite webpages. A good example of this was a list created in 1994 by Jerry Yang and David Filo, grad students at Stanford University. Their website was called *David and Jerry's Guide to the World Wide Web*. It wasn't a simple list, but was cleverly structured as a hierarchy, enabling easy browsing of its content. A year later David and Jerry re-branded their site and launched it as *Yahoo!*

Managing hand-compiled lists or directories of web pages was really inefficient and anyone could see that as the web grew this was unmanageable; automation was required. In 1993 Matthew Grey, an MIT student, coded the first *web robot* or *crawler* called the *World Wide Web Wanderer*. His *bot* (short for robot) would automatically crawl the web following all the links it could find to find new websites. If the bot saw a link on a web page that was not in its *"already seen that"* list it would follow the link, add the page to the *"already seen that"* list and look at any outward links from that page. In this way the bot could crawl the entire web, in theory, adding only pages it hadn't seen to its list. This list was used to generate a search-able index called *Wandex*. Because of time constraints the WWW-Wanderer didn't index all the text of the web pages it visited – just the pages' titles and headings.

WebCrawler was the first full-text crawler-based search engine and this technique became the standard for web search engines in the mid 1990s. *Lycos*, *AltaVista*, *Infoseek*, *Magellan*, *Inktomi*, *Northern Light* and *Excite* all used this approach, and along with the directory-based Yahoo! they quickly became the hottest properties on the stock market. That is, until a small start-up called *Google* changed the way web search was done and became a multi-billion dollar com-pany. But that story is the subject of the next chapter.

Chapter 9

DOTCOM

Whilst at Cornell University, Stephan Paternot and Todd Krizelmann, two young students, had used a chat service on the university computer network. They decided to build their own chat service that would run on the Web. *theGlobe.com* went live in 1995 and attracted over 40,000 users within its first month. In 1997 they raised $20 million in venture capital – they were still both only 23 years old. At the end of 1998 theGlobe.com had its initial public offering (IPO). The company had never made a profit; indeed it's not clear if it had ever earned any money at all. The initial stock soared a record-breaking 606% in its first day of trading – Paternot and Krizelmann were now each worth $100 million!

A year later Paternot was filmed for a CNN documentary about the new dotcom economy. He was out in Manhattan partying in a trendy nightclub. He's clearly drunk, wearing leather pants, and is dancing on a table with his model girl-friend. During the filming he shouts out: "I *got the girl. Got the money. Now I'm ready to live a disgusting, frivolous life*." The following year theGlobe.com's stock price plummeted from a high of $97 to ten cents.

Sir Tim Berners-Lee never made any money directly from inventing the World Wide Web, although he has a very successful career because of it and a knighthood from the Queen. Like many great people in the history of the computer, he was motivated by the general good. Indeed, had there been a charge for using his invention, it is possible it may never have caught on at all. However, many people have made and lost millions on the Web. This chapter is about them.

I. Watson, *The Universal Machine,*   183
DOI 10.1007/978-3-642-28102-0_9,
© Springer-Verlag Berlin Heidelberg 2012

POWER TO THE PEOPLE

The great power of the Web is that anyone with modest skills can create a web page for themselves. If you have access to the Internet and can use a simple word processor, then you have enough to make a basic website. You can make a website about yourself or your family, about your hobbies, or for a club or society, or for your business. Then something magical happens; once you've uploaded your web pages, they instantly become accessible to everyone with access to the Web anywhere on the planet.

A stamp-collecting club in England can view pages about rare stamps in Australia. Political activists in Indonesia can get information from sources in the US. Conservationists can follow Greenpeace's voyages in the Southern Ocean fighting Japanese whalers. A whole universe of information on every topic imaginable is available 24/7. The business community is not excluded from the Web because nobody is excluded – it's free and open to all. Very quickly, companies started to realize they could make money by using the Web. At first this was centered on products and services that students, the early adopters, wanted. Pizza was probably the first product bought online from a *Pizza Hut* franchise in Santa Clara, California.

The beauty of the Web to small business owners was that they could make a website to showcase their products and services and get their business online relatively cheaply in comparison to other means of advertising. Before the Web, if you ran a small business you might take out adverts in your local paper or perhaps a small spot on a local radio station. You could take out an advert in the *Yellow Pages*, you might produce flyers to leave in appropriate places, and you'd attend relevant trade shows and local events. But that was about it; your advertising reach was very local, since national advertising was far too expensive. International advertising was unthinkable.

Let's say it's 1993 and you own a shop specializing in widgets: large, small, new, second-hand and collectable widgets. You are primarily relying on passing traffic and word of mouth to spread your widget business. Plus you may take out small adverts in widget magazines (if you can afford to) and you may attend widget trade-shows and widget fairs. Your most valuable marketing resource is your customer mailing list. Several times a year you send out your catalogue to your customers, but because of space, printing and postage costs you can't afford to list every type of widget you have in stock or afford to post many catalogues overseas. Days sometimes go by and nobody walks into your shop, or phones asking if you have a particular widget they're looking for.

Fast-forward now five years to 1998. You have a website, www.widgets.com, that is attached to a searchable database that lists every widget you have in stock. Featured widgets include details, photos and users' comments. Customers from all over the world can now see your widget store, email you inquiries and arrange to purchase your widgets. You can also proactively email special offers to good customers. Your little widget shop now has a global reach and you're spending less on marketing than you did five years ago. It's not hard to see why businesses love the Web.

SEARCH AND YOU WILL FIND

In an ideal world, if you were looking for a widget using the Web, your search engine would search every single web page to find just those that were about widgets, and then it would sort them so the best widget pages were listed at the top. This sorted list would then be shown to you. The problem is that the web crawlers used by all the main search engines mentioned at the end of the previous chapter have two main failings. First, they can't search every page on the Web because there are just too many,[1] and it may take time for new pages to be crawled and added to their index. Second, the crawlers are dumb; they have no way of telling if a web page about widgets is the best,

---

[1] 12.4 billion indexed web pages on Nov 2 2011 http://www.worldwidewebsize.com

good, bad or really awful. If someone makes a web page titled, *"web sites **not** to visit if you want widgets"* it's as likely to be included in the search results as, *"truly the world's best and cheapest widget site."*

Web searching changed when Larry Page and Sergey Brin, PhD students at Stanford University, realized that they could do searches in a new way. Basically, the crawler-based search engines count the number of times that a search term, let's say *"widget"*, occurs on a page. The more times *"widget"* occurs, the higher the score the page is given.

Website designers knew this, and would fill their pages with keywords to artificially boost their scores. In August 1997, when I heard of the tragic death of Princess Diana, I went online to conduct a simple experiment. I wanted to see how quickly the web would respond to this event. I typed *"princess Diana death"* into my search engine and waited for the results. The top ranked page led me to a porn site! Some cynical person had quickly redesigned the porn site to include words people may be searching for. Sometimes these words were white text on a white background – invisible to you and me, but visible to the web crawlers.

Sergey Brin and Larry Page, Google's co-founders

Page and Brin's ranking method is different. They realized that a *good* web page would probably have more links pointing to it than a *bad* web page, since people will tend to create links to good web pages but not to bad pages. This is similar to the way scientific papers gain a *citation score*; influential papers have a higher score than papers with less influence. So by counting the number of pages that link into a page, they can determine the *PageRank* of a page. Many people incorrectly think that the term *PageRank* is named after Larry Page – it isn't.

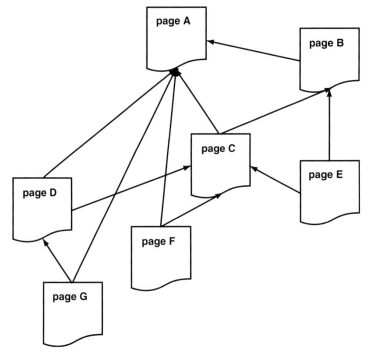

A web of linked pages

If you look at the diagram above you can easily see that "Page A" would have the highest PageRank since it has the most pages (5) linking to it, followed by "C" (3), "B" (2), "D" (1) and "G," "F" and "E" with no links to them. Page and Brin decided to call their search engine *Google,* a misspelling of the word "*googol*" that signifies a one followed by a 100 zeros, since their search engine would have to search huge amounts of information. Google initially ran on computers at the University of Stanford, until they formed a company in 1998 and moved it into a friend's garage in Menlo Park, California.

In 1999 Page and Brin felt that Google was distracting them from their studies and they offered to sell Google to *Excite,* one of the leading search engine companies at the time, for $1 million. In a decision the executives at Excite came to call, "*a stupid business decision,*" they turned the offer down. In 2004 Google's IPO valued the company at $23 billion. Page and Brin were instant multi-billionaires and Excite had missed the deal of the century. Page and Brin never did finish their doctorates, although they have since picked up numerous awards and honors.

Google's *PageRank* technique made the Google search engine so much better than the competitors that once people started using Google they rarely bothered to use any other search engine again. In 2011 Google has almost 90% of the search engine market share while its competitors, like Microsoft's Bing and Yahoo!, are not even close to double figures. To calculate the PageRank of every page on the Web, Google has to crawl the billions of web pages in existence and record all of the links between pages. This is a massive task, and in addition Google has to be able to handle a billion searches daily. Consequently, Google operates massive *server farms* that contain over a million individual computers. These server farms consume so much energy that they are often located near cheap sources of electricity, like hydroelectric power stations.

A Google server farm, The Dalles

Google has branched out from search, and now offers a wide variety of products and services such as *Gmail, Google+, Google Docs, Google Maps, Google Earth, Picasa, Chrome*, and even a smartphone operating system called *Android*. Google is attempting to digitize all of the physical books in existence, though it is meeting some legal resistance, and its *Art Project* is making some of the art in the world's greatest museums available online in *megapixel* quality.[2] Although a massive company, Google tries to stay true to its roots; it is the epitome of the innovative and slightly anarchic start-up culture that has characterized Silicon Valley since the 1960s.

The company slogan is *"Don't be evil"* and at its headquarters, called the *Googleplex*, employees have access to free cafeterias, childcare, hairdressers, gyms and a variety of entertainment options including video games, ping pong, pool and even a grand piano. There is no dress code, and staff, called *Googlers*, are encouraged to spend 20% of their time on projects of their own choosing. Google estimates that half of its new products come from this innovation time. Google seeks to attract the brightest people in the world, and it wants them to have fun.

Google is famous for its April Fools jokes – in 2007 in Gmail it announced *Gmail Paper,* a new service where Google would print out an email and post it to the recipient for you. Its jokes extend to Google search itself; if you type in the word "*anagram*" you'll see it returns: "*do you mean nag a ram,*" which is an anagram of "*anagram*". Type in "*answer to the ultimate question of life, the universe, and everything*" and it returns "*42*", the answer from Douglas Adams' brilliant book *The Hitchhiker's Guide to the Galaxy,* which has always been popular with geeks and nerds. In Google Maps, if you ask for directions from Los Angeles, California to Tokyo, Japan, it directs you up the west coast and then advises you to "*Kayak across the Pacific Ocean.*" It's no surprise that amongst students graduating from my university department, Google is seen as one of the very best companies in the world to work for.

---

[2] http://www.googleartproject.com

EVERYTHING FROM A TO Z

Have you ever felt that you missed out on something? Failed to see that band just before they became huge? Failed to buy into that neighborhood just before prices soared? Failed to buy gold just before the price jumped? In 1995 Jeff Bezos, a Princeton computer science graduate felt that he was in danger of missing the Web bandwagon, so in what he called *"regret minimization,"* he decided to do something about it. On a cross-country drive with his wife from New York to Seattle, he worked out a plan for an online bookstore he would call *Amazon* – after the river, because it's big, and because *"A"* will appear first in most lists. The company logo shows a curved arrow from A to Z, indicating the store stocks everything from A to Z, while the curve looks like a happy smile.

Bezos realized that even the largest physical bookstore might only stock 100,000 books. A virtual *e-bookstore* could stock every publishers' entire catalogue. Amazon would be open 24/7 to anyone with a web browser. He also saw that since Amazon wouldn't need expensive main street retail stores and staff, his overheads would be lower than a brick-and-mortar bookseller. Starting in his garage, *"the world's largest bookstore"* was launched and in July 1995 Bezos sold his first book, the fascinating bestseller: *Fluid Concepts and Creative Analogies: Computer Models of the Fundamental Mechanisms of Thought.*

Jeff Bezos, the founder of Amazon.com

Amazon had an unusual business model, since Bezos did not expect the company to show a profit for four or five years at least. It would concentrate on expanding its customer base and diversifying into selling CDs, DVDs, electronics and eventually almost everything. In 1997, Amazon had its IPO of stock on the NASDAQ exchange. The bookseller *Barnes and Noble* promptly sued, asserting that Amazon's claim to be the world's largest bookstore was false, since it was a *book broker*, not a bookstore. The case was settled out of court.

Bezos' business model proved to be correct and Amazon didn't turn a profit until the end of 2001, making just $5 million on a turnover of over $1 billion. However, Amazon had built a loyal customer base, proved that online retailing was viable, and created an important innovation: a personalized shopping experience. When you walk into a physical shop, let's say a bookstore, you'll see well-presented promotions. These will be of the latest book releases, probably the 10 or 20 top-selling books, plus some other promotions; perhaps a politician's memoirs or a biography about a celebrity who is visiting your town, perhaps a special collection of books for Mothers Day. When somebody else walks into the same store they will see exactly the same promotions and displays. Every few weeks the store manager and staff will change the displays – a time-consuming task.

In a virtual store, though, the featured promotions can be personalized for each shopper. Let's assume that I like fishing and you like opera. When I enter Amazon it knows my past purchases and my browsing history, the books and DVDs I look at in detail. The content I see is tailored individually and on the fly to show me DVDs and books about fishing and even fishing equipment is promoted. When you enter Amazon, it knows you like opera and so will reconfigure the store to feature opera content personalized for you: albums, DVDs and books about opera singers. It's like we walked into two separate bookstores, one that specializes in fishing and the other opera.

Amazon uses sophisticated artificial intelligence and machine learning algorithms to compare each customer's history and preferences with other customers so it can make personal recommendations, just as a knowledgeable shopkeeper might. You will have seen on Amazon's pages product recommendations like: *"Frequently Bought Together"* and *"Customers Who Bought This Item Also Bought..."* Amazon is intelligently trying to persuade you to buy more.

You may be reading this book on your Kindle or iPad. If you are one of those people who have traded the comfort of a traditional book for the convenience of e-books you'll not be surprised that in January 2011 Amazon announced in a press release it was now selling more e-books than traditional paperbacks.

"Amazon.com is now selling more Kindle books than paperback books. Since the beginning of the year, for every 100 paperback books Amazon has sold, the company has sold 115 Kindle books. Additionally, during this same time period the company has sold three times as many Kindle books as hardcover books. This is across Amazon.com's entire US book business and includes sales of books where

there is no Kindle edition. Free Kindle books are excluded and if included would make the numbers even higher."

Johannes Gutenberg invented the printing press in 1436. This technological breakthrough enabled books, for the first time, to be cheaply mass-produced. Ever since books have remained central to our civilization. After nearly 600 years the universal machine is changing how books are made, distributed and read.

I COLLECT BROKEN LASER POINTERS!

Another computer science graduate, French-born Iranian American Pierre Omidyar, was also quick to jump on the Web bandwagon. On Labor Day 1995 he launched *Auction Web,* which was later rebranded as *eBay*. A broken laser pointer was the first item sold for $14.83. Omidyar contacted the winning bidder to check, *"Do you know the laser pointer is broken?"* The buyer replied: *"I collect broken laser pointers,"* and so a phenomenon was born. People have always liked auctions; you never know what you'll stumble across for sale, and you may find a bargain. The Web allowed the virtual auction to be colossal and eBay quickly grew, hosting 250,000 auctions in 1996 and two million in 1997. eBay listed on the NASDAQ in September 1998 and Omidyar became another instant dotcom billionaire.

Pierre Omidyar founder of eBay

Anything can be listed on eBay, providing it is not illegal; antiques and collectibles, computers, cars, boats, toys, clothes, services like dog walking and house painting. Companies and businesses can list their products and services, and items can be auctioned for charity. New Zealand, where I live, was once listed by an Australian for an opening price of one cent, whilst the US Virgin Island *Thatch Cay* was listed on eBay in 2003 for an opening bid of $3 million.

eBay taps into our desire to hunt down bargains and to see how much we might make from selling our own junk. *"One person's junk is another's treasure"* certainly seems to apply. eBay quickly spread worldwide through opening eBay sites in other countries and by aggressively acquiring competitors. For example, eBay bought *PayPal* because 50% of its customers were using the service rather than its own *Billpoint* system. With a strong business model and cash flooding in from the fees it generates from every single sale, eBay now employs over 17,000 people globally and has revenues of over $9 billion.

IT'S GOOD TO SHARE

Not all dotcom start-ups were about buying and selling, and many were certainly not online versions of traditional shops or services like auctions. Some start-ups made activities that had once been purely local become global. One such was *Napster*.

I've always liked music, and enjoy buying albums and going to watch bands perform. When I was young, like most kids I didn't have much money, and so there was no way I could buy all the albums I wanted. My friends were all in a similar situation. Because we were friends we had similar music taste, but we never bought the same albums. We'd realized there was a solution to our relative poverty: if we each bought a different album, then we could record onto cassette tapes each others' purchases. Cassette tapes were much cheaper than albums and though the sound quality wasn't as good, it was acceptable. So we'd each spend hours recording each others' music – technically we were breaking the law, but everyone did it.

When I went to university I made new friends and my access to new music expanded. Now it was difficult to find the time to record all the music I had available to me; remember it took 40 minutes to record a 40-minutes long album. CDs were invented, but the cassette tape was still used to record them until computers started to appear with read-writable CD drives in them. Now it was possible to copy a whole CD and get a digitally perfect copy. Groups of friends were still *sharing* their music collections by allowing each other to copy their new purchases.

In 1995 a file standard, known as MP3, was released that supported the compression of digital audio, such as a CD track, into a relatively small file. There was a loss in audio quality, but for most listeners the sound was fine. It was now possible to *rip* the music off a CD and store it as MP3 files on a computer. The computer could then be used to play the music directly. MP3 files could be copied in seconds, so for the first time a music CD, once it had been *ripped*, could be copied in a minute or so. Students now started sharing their music collections as MP3s, and moreover they didn't have to even buy cassette tapes. You could copy a friend's entire music library in minutes at no cost.

Shawn Fanning co-founder of Napster

You were still limited to your circle of friends, but the Web was about to make your circle go global. Shawn Fanning, a student at Northeastern University in Boston, wrote a program that allowed people to share MP3 files over the Web. Launched in June 1999, with his uncle John Fanning as a major investor, it was called *Napster* after his nickname. Just as I shared my music with my friends (my *peers*), Napster was a *peer-to-peer* file sharing service. You could list the music you wanted to share and see the music everyone else was sharing. The beauty of Napster was that it was easy to use, but also that everyone's circle of friends now included all Napster's users. In less than two years there were 20 million users and 80 million songs were listed. Sean Parker, who will appear

196

later in this story, is sometimes attributed as being a co-founder of Napster, though there is confusion as he may have just been an early employee. Parker realized the power of connecting people together, in Napster's case through a love of the same music, and later in our story with Facebook.

Napster overloaded the networks of colleges around the US, with some estimating that over 60% of external network traffic into and out of campuses was Napster traffic. Needless to say, many colleges attempted to block Napster and made its use a disciplinary offence. But colleges were not the only ones taking offence.

The music industry immediately took notice. It had never liked it when people recorded music onto cassette tapes, but had been unable to do much about the common practice except attempt to call on peoples' sense of decency and honesty. Now, instead of thousands of schools kids making a few cassette tapes, they had a single company that was making it possible for people to illegally share their products on a colossal and global scale. In December 1999, a few months after Napster's launch, A&M Records and several other recording companies filed *A&M Records, Inc. vs. Napster, Inc.* claiming that Napster was infringing their copyright.

As the case dragged through the courts, Napster became more and more popular, ironically because of the publicity generated by the legal case. In 2001 the popular heavy metal band *Metallica* discovered that an unreleased demo of a song called "*I Disappear*" was available on Napster. The song was even played on the radio before they had released it! They then discovered to their horror that their entire back catalogue was available to download from Napster. You could be a Metallica fan and own every one of their albums, including several live bootlegs, without ever having to pay a single cent. Metallica sued Napster and were quickly followed into court by the rapper Dr. Dre, who had the same lawyer. Napster settled out of court. Napster also lost the case against A&M and in 2002 filed for bankruptcy.

THE DOTCOM BUBBLE

Not every dotcom company had a strong business model; indeed some didn't seem to have a business model at all. Some of the start-ups coming out of garages and university labs had breakthrough technology, like Google, or a solid business plan like eBay; realistic plans for slow, steady, planned growth that would eventually lead to profits, like Amazon; or even crazy exponential growth fueled by college kids getting something they wanted for nothing, as with Napster. However, the soaring stock market valuations of the early dotcom companies encouraged investors to believe that this was a "*new economy*" and that any company with a ".*com*" after its name or an "*e-*" prefix (for *e-business*) couldn't fail.

Instead of showing how they would make a profit from their website, start-ups were convinced that their eventual huge number of users would in some way magically make them profitable. They would suck up large initial losses until the *network effect* kicked in and their sector dominance made them billionaires. There were two serious problems with this thinking. First, only one dotcom could dominate each market sector, like Google in search and Amazon in retailing, so there would inevitably be many losers in each sector. Second, nobody actually knew how to turn eyeballs into money. It was assumed that targeted adverts and *click-through* payments would somehow generate a profit. The key to success was growth, a catchy name, growth, lots of publicity (good or bad), and more growth.

The public offerings of companies like eBay and their rapidly soaring stock prices, combined with low interest rates, drove investor enthusiasm. Changes in investment and stock trading practices also fuelled the demand. In 1997, the US *Securities and Exchange Commission* implemented the *Small Order Execution System*, which required market traders to buy or sell small orders (up to 1,000 shares) instantly, rather than waiting a week to settle. This and other technological advances encouraged a new breed of *day-traders* into the stock market.

Day-traders were often not financial professionals, but ordinary people who had quit their regular jobs to trade on the stock market, taking their profits at the end of each day – hence the name. Working from home, they could obtain up-to-the-minute market data online and could place their trades using newly developed software applications. These technological and legislative innovations coincided with a *bull market* in tech stocks at the end of the 1990s. A bull market is where stock prices rise continually, as opposed to a *bear market,* where they fall. Clearly it is easy for the inexperienced to make money in a bull market, but only the skillful can make a profit in a bear market. From 1997 to 2000, the NASDAQ (the main market for tech stocks) rose from 1200 to 5000. Many new day-traders, with little market experience, made huge profits buying these stocks in the morning and selling them in the afternoon.

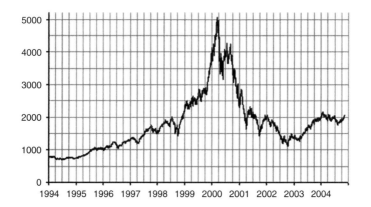

Average NASDAQ share prices between 1994 and 2004

In March 2000 the bubble burst; reason prevailed and people started to see that the underlying finances of many of the dotcom companies made no sense. The story of *boo.com* is illustrative: founded in 1998 by Ernst Malmsten and Kajsa Leander they wanted to create an online store where the cool and chic would buy clothes – boo.com was to be the Amazon of fashion. They raised millions from some of the biggest banks and investors in the world. They bought fashionable headquarters on London's Carnaby Street, spent millions on advertising, hosted lavish launch parties around the world, and paid themselves generous salaries before they had ever sold a single garment. This in the business was known as the *burn rate* – the speed at which a start-up would burn through its investors' money before it turned a profit. If the burn rate were too high, the company would run out of money before it started making a profit. Boo.com burned through £125 million in six months; it sold just £200,000 of clothes in its final two months and went back to the market for a further £30 million. But by now the bubble had burst and no investors were willing to put their money into a company that had never turned a profit. Boo.com was bankrupt.

The day-traders and people like Stephen Paternot, Todd Krizelmann, Ernst Malmsten and Kajsa Leander lost all their money and learned the hard way that, as it says in the fine print, "*investments can go up as well as down.*" Dotcom had become *dotbomb* and many of the start-ups were now insolvent. It would be a decade before optimism would return to the tech market, but that is a story for another chapter – first the universal machine must become so small it fits in your hand.

Chapter 10

THE SECOND COMING

*I*t's Sunday October 16 2011 and you're inside the beautiful Memorial Church at Stanford University, which glitters like the inside of an exquisite jewel box. The church is busy, filled with political, business and technology leaders plus a clutch of actors and musicians. Former president Bill Clinton and his family and former vice-president Al Gore are here, along with Jerry Brown, the Governor of California and former first lady of California, Maria Shriver; Microsoft's Bill Gates, Google's Larry Page and Sergey Brin, News Corp CEO Rupert Murdoch, Oracle CEO Larry Ellison, Michael Dell of the eponymous computer company, Adobe co-founders John Warnock and Chuck Geschke, Samsung's Jay Lee and a myriad other tech industry figures. The actor Tim Allen and Pixar's creative genius John Lasseter are here sitting by cellist Yo-Yo Ma Bono from U2 and the British writer, comedian and geek, Steven Fry. Joan Baez, the folk singer, is singing, "*Swing Low, Sweet Chariot.*"

You've all come on a clear cool Sunday evening to celebrate the life of a remarkable man—Steve Jobs of Apple who died last week of pancreatic cancer, age 56. As you listen to the tributes you recall the *Commencement Address*, which Jobs gave to Stanford graduates in 2005. It seems so apt now:

"No one wants to die. Even people who want to go to heaven don't want to die to get there. And yet death is the destination we all share. No one has ever escaped it. And that is as it should be, because Death is very likely the single best invention of Life. It is Life's change agent. It clears out the old to make way for the new. Right now the new is you, but someday not too long from now, you will gradually become

I. Watson, *The Universal Machine*,
DOI 10.1007/978-3-642-28102-0_10,
© Springer-Verlag Berlin Heidelberg 2012

the old and be cleared away. Sorry to be so dramatic, but it is quite true."

—Steve Jobs

Video of the address,[1] which ends with his advice to: "*Stay Hungry. Stay Foolish*" has gone viral on the Web and people all over the world are mourning his death. You've never seen anything like it, for John Lennon, Princess Diana, Michael Jackson sure, but not for a businessman! Over a million people have left messages on a tribute page at Apple.com and people are leaving flowers at makeshift shrines outside Apple stores all round the world. The word *genius* and *visionary* appear in almost every obituary and news channels have been airing wall-to-wall *special features* on Jobs for days now. The story of how Jobs brought Apple back from the brink of bankruptcy and in the view of some, "*changed the world,*" is the subject of this chapter.

STAY HUNGRY, STAY FOOLISH

We left Steve Jobs back in chapter seven, with Woz taking leave from Apple after his plane crash. Jobs' story is just as interesting as Woz's, so let's backtrack to his childhood. Steve Jobs was born February 24, 1955 to an unmarried graduate student in San Francisco, Joanne Simpson, who gave him up for adoption. Paul and Clara Jobs, from Mountain View, California, adopted the baby and named him Steven. Jobs' biological mother later gave birth to a daughter called Mona Simpson. She went on to marry a TV scriptwriter, Richard Appel, who wrote for the hit cartoon series *The Simpsons*. Appel wrote the episode in which Homer Simpson's mother returns; she is called Mona Simpson, after Appel's wife and Jobs' biological sister. So in weird way, Homer Simpson is Steve Jobs' nephew!

Jobs attended Homestead High School in Cupertino, California along with Woz, although Jobs was four years his

---

[1] http://youtu.be/D1R-jKKp3NA

junior. Like Woz and many of the kids in the neighborhood, Jobs had a keen interest in electronics and liked building gadgets. He was a member of his high school electronics club, calling themselves the *Wireheads*. In Silicon Valley it was cool, not nerdy, to be into electronics. While Jobs didn't have Woz's genius for creating circuits, he did have another skill just a valuable—chutzpah. Once when the Wireheads needed some parts, Jobs phoned Bill Hewlett, the co-founder of Hewlett-Packard. "*He was listed in the Palo Alto phone book,*" said Jobs. "*He answered the phone, and he was real nice. He chatted with me for, like, 20 minutes. He didn't know me at all, but he ended up giving me some parts.*" Jobs was a natural deal maker and perhaps he felt he knew Bill Hewlett since he was a member of the *Hewlett-Packard Explorers Club*, a small group of kids who met once a week to listen to HP engineers talk about their work.

Woz was working at Hewlett-Packard after he dropped out of Berkley. Jobs got a summer job there, and it was around this time the two Steves started making and selling their blue boxes for phone phreaking. Jobs said in his biography, "*If it hadn't been for the blue boxes, there wouldn't have been an Apple, I'm 100% sure of that. Woz and I learned how to work together, and we gained confidence that we could solve technical problems and actually put something into production.*" In 1972, Jobs graduated high school and enrolled in the liberal arts Reed College in Portland, Oregon. He dropped out after just one semester because he couldn't justify his parents spending their life savings on his college tuition. He continued attending classes without paying, such as one on calligraphy, whilst he slept at friends' homes and got free meals at a local Hare Krishna temple. Jobs has since said, "*If I had never dropped in on that single course in college, the Mac would have never had multiple typefaces or proportionally spaced fonts.*"

In 1974 he returned to Silicon Valley, re-established his friendship with Woz and started attending meetings of the Homebrew Computer Club. Jobs got a job at the new computer games start-up *Atari* because he wanted to save some money to go on the hippie trail in India to find enlightenment.

Jobs wasn't exactly an ideal employee; he believed that his vegetarian diet prevented body odor even if he didn't wash or use deodorant. To be blunt, Jobs smelt so bad other employees complained; Atari's boss Nolan Bushnell, who liked Jobs, solved the problem by having Jobs work alone at night.

Jobs left Atari to go to India, but was disillusioned by the poverty and the filth he found there. Nonetheless he became a Buddhist, shaved his head, and was wearing robes when he returned to California. His vegetarianism became more extreme and he experimented with strange diets, like just fruit, and purges. It was at this time that Jobs, like many others in the Bay Area, experimented with LSD, which he says, *"was a profound experience, one of the most important things in my life. LSD shows you that there's another side to the coin...creating great things instead of making money."*

Steve Jobs' family home in Los Altos, California. Apple computer was founded in its garage

As we read in chapter seven, he helped Woz build and sell the Apple I, but had very little input into its design; it was really all Woz's creation. So when Woz started work on the Apple II, Jobs took responsibility for the external design of the product. Jobs didn't understand why a computer had to look like a bulky, ugly piece of electronics, as if a ham radio enthusiast had built it in his garage. He wanted the Apple II to look like a chic piece of consumer electronics, like a European hi-fi. So he set to work designing the external case of the Apple II—a smooth, sleek design totally unlike other computers of the time. For its launch Apple's logo was redesigned; Ron Wayne's Victorian etching was ditched in favor of a simple rainbow hued apple with a bite taken out of it. The Apple II brochure featured a simple quote attributed to Leonardo da Vinci, "*Simplicity is the ultimate sophistication.*" This would become Jobs' mantra.

The Apple II was a sensation when it was unveiled at the *West Coast Computer Faire* in 1977, and the young company struggled to keep up with demand. Jobs for a while even considered selling the Apple II to Commodore. He named a price: $100,000, generous amounts of Commodore stock, and salaries of $36,000 for both Woz and himself. The deal fell through, so we'll never know what may have happened to Jobs and Woz if Commodore had made them employees. In 1979 Apple had its IPO and made more millionaires overnight than any company in history. The Apple II remained on sale for 16 years and sold nearly six million units, more than any other personal computer.

We already learnt in chapter seven how Jobs had cheated Woz out of his fair share of a bonus whilst he was working for Atari. There is another example of Jobs' strange way of treating people close to him from this time. For several years Jobs had been dating a young woman called Chrisann. She became pregnant and nobody doubted that Jobs was the father—that is, nobody except Jobs himself. For years Jobs denied it and refused to have anything to do with the child, called Lisa, even when a DNA paternity test showed that Jobs was almost certainly the father and a California court ordered him to pay child support. Eventually, and it seems because of a fear of the bad press it might bring Apple, Jobs settled a relatively small sum of money on Chrisann, even though he was now worth $100 million dollars.

You may think Jobs was a greedy man, willing to do-over his friends and acquaintances for a few dollars, despite being a Buddhist. Certainly he was not known for his philanthropy, unlike Bill Gates and Woz. Yet, paradoxically, Jobs wasn't really that interested in money. He always wore the same simple and inexpensive clothes, lived in a relatively modest Palo Alto home, considering his wealth, and although he loved German sports cars, he just drove a late model Mercedes. Curiously, his car had no number plates.

In 1982 Jobs bought an expensive two-story apartment in Manhattan and spent a small fortune refurbishing it, never to move in. He sold it to his friend Bono from U2. In 1984 Jobs bought a huge Spanish colonial mansion in California originally built for a mining baron. Jobs lived there for many years but never refurbished it; indeed he never even furnished it. Rumor has it that Jobs slept on a mattress on the floor and in the main living room there were just a couple of hard chairs and his BMW motorbike. Was this perhaps a Zen-inspired desire for simplicity or just because he couldn't be bothered to spend his time buying furniture?

Another aspect of Jobs that can't go unmentioned was his unpredictable management style. He had an ability to foster great loyalty and commitment from people. In 1981 Bud Tribble, an Apple employee coined the phrase, "*Reality Distortion Field,*" to describe the way Jobs could make people around him believe and commit to the seemingly impossible. There is no doubt that Jobs could be very charismatic, but equally he didn't tolerate fools. Apple employees during the bad times tell how they avoided using the elevators, preferring to take the stairs, in case they got caught in the elevator with Jobs. It was called being "*Steved.*" Allegedly if you were trapped in the elevator with him he might ask you a question like: "*What do you do here and why are you worth what I'm paying you?*" If you gave an answer he didn't approve of he'd fire you on the spot. Conversely though, others say Jobs relished dissent; if you disagreed with him and stuck to your guns, arguing your position rationally, the meeting would get very heated and Jobs might lose his temper. However, he would eventually concede he was wrong and respect you all the more.

CHANGE THE WORLD

"The best way to predict the future is to invent it."
—Alan Kay, Xerox PARC

In 1979 Jobs along with several other Apple employees visited Xerox PARC to see what they had been working on. Remember, Xerox was an early investor in Apple and was keen to show Apple its Alto computer system, its graphical user interface and other developments at Xerox, such as networking and object-oriented programming. Jobs has said about his visit: *"I thought it [the GUI] was the best thing I'd ever seen in my life… within ten minutes it was obvious to me all computers would work like this…it was like a veil being lifted from my eyes. I could see what the future of computing was destined to be"* He has also observed that: *"Picasso had a saying: Good artists copy, great artists steal."* Jobs always acknowledged that Apple took ideas from Xerox. He left Xerox that day determined to be the first to bring the GUI to the market.

At this time Jobs was working on a successor to the ill-fated Apple III called Lisa. Yes, that's right, the computer had the same name as the little girl Jobs refused to acknowledge as his daughter. However, he quickly realized that Lisa would not have enough processing power to run a GUI, and so he set about gathering the best Apple talent (and some taken from Xerox) into a new project called Macintosh. In a set of rooms sequestered away from the rest of the company, the Macintosh team set about designing its hardware and software. Once again, Jobs had very firm ideas on what the product should look like. At one meeting he threw a phonebook on the table and said: *"That's how big the Macintosh can be. Nothing any bigger will make it. Consumers won't stand for it if it's any larger."* When the design was complete, he insisted that the team sign the inside of the Macintosh case, just like artists might.

Just before the release of the Macintosh, Mike Markkula, an early angel investor in Apple and its President, decided to retire. He and Jobs agreed that a professional replacement was required to oversee the day-to-day running of what was now a *Fortune 500* company. Jobs, in a famous pitch, persuaded the President of *Pepsi-Cola*, John Scully, to leave and join Apple, challenging him if he preferred to: *"sell sugar water for the rest of your life or come with me and change the world?"* Jobs was now free to concentrate on finalizing the Macintosh.

After the hugely successful Super Bowl commercial, Jobs revealed the Macintosh to the public in his now-classic theatrical style. He lifted a Macintosh out of a small bag. Using a new experimental voice synthesis program, it said to the audience:

"Hello, I am Macintosh. It sure is great to get out of that bag. Unaccustomed as I am to public speaking, I'd like to share with you a maxim I thought of the first time I met an IBM mainframe: never trust a computer you can't lift! Obviously, I can talk, but right now I'd like to sit back and listen. So it is with considerable pride that I introduce the man who's been like a father to me"

Jobs' insistence on the clean, simple look of the Macintosh had produced an aesthetically pleasing machine, but it had come at a cost. The limitations of the first Macintosh soon became obvious: it had very little memory compared to other PCs in 1984, it could not be expanded easily, which had been the great strength of the Apple II, and it lacked a hard disk drive or the means to attach one. Apple started a scheme where prospective buyers could take a Macintosh home for 24 hours and just leave their credit card details with the dealer. However, this was unpopular with the dealers, and expensive, because returned trial machines could not be resold. This campaign caused John Sculley to raise the price from $1,995 to $2,495; that's about $5,200 in today's money. Macintoshes were expensive and despite their groundbreaking design and revolutionary GUI, they weren't selling well and Apple was in increasing financial difficulty. The Apple III had been a commercial flop; it hadn't the brilliant design input of Woz and worst of all they were manufactured poorly and constantly broke down. The old Apple II still created all of Apple's profits eight years after its release.

In 1985 a power struggle developed between Jobs, John Sculley and the Apple board. Jobs' obsession with experimental products, and his own belief that he knew what people wanted better than they did, was seen as a weakness and not a strength. Jobs once said *"We're gambling on our vision, and we would rather do that than make 'me too' products. Let some other companies do that. For us, it's always the next dream."* He has also been quoted saying: *"It's really hard to design products by focus groups. A lot of times, people don't know what they want until you show it to them."*

Jobs got wind of Sculley's boardroom maneuverings and attempted to oust him from his leadership role at Apple. But Sculley found out that Jobs was organizing a putsch, and called a board meeting where Apple's board sided with him and removed Jobs from all his managerial duties. Although still an Apple employee and a major stockholder, Jobs now had no official role at Apple. He was just a few days off his thirtieth birthday.

In an interview after leaving Apple, Jobs said: *"What can I say? I hired the wrong guy. He [John Sculley] destroyed everything I spent ten years working for, starting with me, but that wasn't the saddest part. I'd have gladly left Apple, had Apple turned out like I'd wanted it to."* Just ten years after Woz and Jobs had started Apple, neither was now actively involved with the company, but Jobs' involvement with computers and Apple was not over yet.

WHAT'S NEXT

"I feel like somebody just punched me in the stomach and knocked all my wind out. I'm only 30 years old and I want to have a chance to continue creating things. I know I've got at least one more great computer in me. And Apple is not going to give me a chance to do that."
— Steve Jobs in a *Playboy* interview

Before he was ousted from Apple, Jobs had met Paul Berg, a Nobel Prize-winning chemist, who outlined his frustration that affordable high-powered computers were not available for his students. So when Jobs left Apple, he took a few good people with him, mostly from the Macintosh team, and started a new company to design and manufacture a computer that would be ideal for the university market. The company, which he called *NeXT Inc.*, was to develop a powerful computer intended for scientists and students. Many aspects of the NeXT workstation were innovative; from its radical black mag-nesium cube-shaped design, to its built-in Ethernet connec-tivity and object-oriented operating system called NeXTSTEP (two technologies Jobs had seen at Xerox PARC). Ross Perot, the one-time independent US Presidential candidate, and *Canon* became large investors in NeXT.

Jobs created an unusual flat corporate management struc-ture at NeXT, with all employees on just two salary scales: $75,000 and $50,000 per year dependent on when they were hired. This resulted in some staff earning more than their managers before bonuses. The company payroll, unusually, was open for all employees to inspect, and to further encour-age transparency, NeXT's offices in Palo Alto were entirely open plan; only Jobs had an enclosed office. Perhaps Jobs was trying to ensure that NeXT would not encourage the secret scheming and conspiracies that had so troubled him at Apple.

His troubles with Apple were not over though; before the design of the NeXT workstation (or *the Cube* as people called it) was finalized, Apple sued claiming Jobs was using its tech-nology. Jobs commented, *"It is hard to think that a $2 billion company with 4,300-plus people couldn't compete with six* *people in blue jeans.*" The case was dismissed before it came to trial.

Tim Berners-Lee's NeXT computer at CERN—the first web server

A state-of-the-art manufacturing facility, inspired by Japanese *just-in-time* manufacturing methods, was established and the first computers were sold for $6,500 in 1989. Although this seems expensive, these were very powerful machines for their day; an order of magnitude more powerful than a PC of the time. However, there was a new problem: competitors like *Sun Microsystems* were now almost giving their workstations away to university departments in return for lucrative software licensing deals. NeXT found it very hard to compete against this practice. Newer, more powerful versions of the NeXT workstations were developed, including one that featured the new CD-ROM drive. By 1992 around 50,000 NeXT computers had been sold, but Jobs took the difficult decision to pull the plug on the loss-making hardware business, and two-thirds of the 500-plus workforce were made redundant.

However, the software NeXT developed was much more successful and has been very influential, if you use any modern operating system: Windows, Linux or OS X elements of NeXTSTEP, like its dock, are now a common feature. Indeed the Mac's OS X derives its code-base directly from NeXTSTEP, but more on that later. NeXTSTEP is based upon a UNIX core, derived from *AT&T Bell Laboratories* and used on many commercial mini and mainframe computers. Because most of these computers handle time-sharing and multiple user accounts, UNIX has security access and file privileges built in from the ground up. This makes UNIX and its variants very secure systems, and less prone to viruses and other malware that plagues Windows systems that were originally designed for single users.

NeXTSTEP was also notable for its complete use of *object-orientation*, a new programming paradigm that dramatically increases the modularity and reliability of software, and the productivity of programmers. Before object-orientation, computer software was written and executed like a musical score. The composer (the programmer) wrote the music (the code) from the start to its finish in a continuous linear stream. In object-orientation a computer program is made up of a collection of modules or *objects*. Each object is independent but can communicate and link with other objects. It's as if a composer composed by saying, "***start** using **quiet introduction**, then **drum roll**, then **melody**, repeat **melody**, then **crescendo** then **end**.*" Each of the components were pre-written small musical parts that could be easily combined and altered.

Mozart might not like this, but less gifted musicians and most computer programmers would make music much quicker. They might even make quite good music if a talented musician, like Mozart, composed the individual musical parts. In essence this is how object-oriented programming works. The majority of modern computer programming languages are now object-oriented and once again the original concept came from Xerox PARC, invented by Alan Kay. NeXTSTEP provided several *toolkits,* or *kits* as programmers call them, which were used to build all of the software on the NeXTs. These were very powerful and they made the writing of applications with NeXTSTEP far easier than on other computer systems. Tim Berners-Lee used a NeXT at CERN to create his first web server and website, and NeXTSTEP was considered a paragon of computer development for years to come.

SOME LIGHT ENTERTAINMENT

"I think Pixar has the opportunity to be the next Disney – not replace Disney - but be the next Disney."
—Steve Jobs

NeXT wasn't Jobs' only business venture after leaving Apple; he also bought a small computer graphics company. In 1979 a team of academics, led by Dr. Ed Catmull from the *New York Institute of Technology*, had developed some new computer graphics techniques and went to work for George Lucas, the director of *Star Wars*. They formed the *Graphics Group* of *Lucasfilm,* working on special effects for *Industrial Light and Magic,* creating computer-generated scenes like the *Genesis Effect* in *Star Trek II: The Wrath of Khan*.

Real estate agents say that "*debt, divorce and death*" eventually force the sale of even the most desirable properties, and this proved to be true for George Lucas. He and his wife divorced, and under Californian law he had to settle half of his assets on his wife. At the time he was strapped for cash because of the box office flop of *Howard the Duck* (no, I'd never heard of that movie either) and he was forced to sell the Graphics Group for a knockdown price. Steve Jobs bought it in 1986 for a bargain $5 million cash plus a further $5 million capital injection into the company. He renamed the company *Pixar*.

Jobs didn't really buy the company because of the work it did in computer animation. He was interested in the hardware that it had developed to enable it to create its digital magic. Making realistic computer-generated images uses massive amounts of computing power. Imagine a three-dimensional model of a room illuminated by a single light source: suppose the room contains furniture and other objects. To create a lifelike image from a particular viewpoint you have to calculate how an imaginary ray of light will travel from the light source and reflect off every visible point of every surface of everything in the room to the viewpoint. You also have to calculate how reflections may occur on shiny surfaces in the room, and how shadows may fall, and even how light may pass through translucent materials. Now as you move through or around the room, you have to recalculate everything for each new viewpoint on your route.

This is called *ray tracing,* and is the basis of all computer graphics. The process of calculating how all the light will be reflected and viewed is called *rendering,* and it is very computationally intensive. For this reason the scientists at the Graphics Group designed and built their own very powerful computers. It was these powerful computers and the software that they had created for image manipulation that Jobs was really interested in, not the animations they made.

A P-II Pixar Image computer

Pixar tried to sell the powerful *Pixar Image Computer* to hospitals for medical imaging and to the government for image processing tasks. However, apart from a few machines sold to the CIA for processing photos from spy satellites, sales met with little success and Jobs was forced to pour more and more money into the loss making company.

At the time a Pixar employee, called John Lasseter, was making short animated films to showcase Pixar's advanced techniques. These were shown at *SIGGRAPH,* an academic conference for computer graphics, and Pixar and Lasseter quickly gained a reputation for doing revolutionary work. One short film shown in 1986, called *Luxo Jr.,* amazed everyone. The computer animation was a technical triumph, but more importantly, people were able to empathize with the two characters in the short film, even though they were at face value just *Anglepoise* desk lamps. Lasseter already knew, and Jobs then realized, that it wasn't the hardware or the software that really counted; it was the experience that the audience had that mattered.

One company that had bought a Pixar Image computer was *Disney,* which was trying to improve the efficiency of its laborious pen, ink and hand-coloring animation process. Even after Jobs had been forced to sell off the loss-making Pixar hardware division, he maintained a relationship with Disney and in 1991 he struck a remarkable deal with them; they would pay Pixar $26 million for three full-length animated feature films. Jobs thought he had the best side of the deal because the films would be released under the Pixar brand, not Disney, and Pixar would have complete creative control. However, what he didn't realize was that $26 million was a bargain price for three feature-length movies. Jobs, usually a skilled negotiator, was naive when it came to Hollywood. Nonetheless he had the last laugh.

Unusually for Jobs, he was very hands-off in his management of Pixar, perhaps because he recognized he knew little about animation or movie making, or perhaps because he was busy with NeXT. For whatever reason, he mostly left them to it; he certainly didn't micro-manage Pixar. The first movie it released was *Toy Story*. Pixar's IPO had been scheduled to take place one week after *Tory Story* opened. This was big gamble on Jobs' part since if the movie bombed, who would want to buy a stake in the company that made flops? Jobs' gamble paid of as *Toy Story* became the top-grossing movie of all time in its first weekend, going on to earn over $360 million worldwide. It also won numerous critical accolades and is now considered a pioneering animated film.

The IPO was a huge success and Jobs' stock in the company was now worth $1.2 billion and Pixar had the money to make many more animated movies. Since then all the films produced by Pixar (the *Toy Story* series, *Monsters Inc.*, *The Incredibles*, *WALL-E*, *Finding Nemo*, *Cars*, etc.) are among the top 50 grossing films of all time. In 2006 Disney, unable to compete creatively with Pixar, bought the company for $7.4 billion dollars. Steve Jobs became Disney's largest shareholder.

ACT II

Question to Michael Dell of *Dell Computers* October 6, 1997: *"What would you do if you owned Apple Computers?"* Reply: *"I'd shut it down and give the money back to the shareholders."*

After Jobs left Apple in 1985, the company hadn't prospered. It had released numerous new models, some targeted at power-users, others for the budget conscious and some for people in between. The *Performa,* for example, was available in a bewildering range of hardware and software configurations that neither sales staff nor consumers understood. Apple had unsuccessful forays into digital cameras, music players, video machines, TVs and even an early personal digital assistant called the Newton. The public was confused. Apple had stood for one thing, the Macintosh, and easy-to-use software, and now the company seemed lost.

In 1996 a new CEO, Gil Amelio, arrived at Apple. It's rumored that when he arrived the company only had enough cash to pay staff their salaries for three months. He recognized a moribund company when he saw one, and he instigated massive layoffs. He also realized that developing a new operating system was crucial to Apple becoming competitive again. After trying to develop an operating system in-house, Amelio went out in 1996 and bought the best operating system he could find—Steve Job's NeXTSTEP. Jobs was brought back in to Apple to act as an advisor. The following year, the Apple board sacked Amelio and Jobs was appointed interim CEO.

Steve Jobs' return to the company in 1996 turned out to be the greatest comeback in business history. Eric Schmidt, Google's chief executive, recently said: *"Apple is engaged in probably the most remarkable second act ever seen in technology."* Jobs' new strategy radically weeded out the plethora of unsuccessful products to focus the company on just four computers: two notebooks and two desktops, aimed at either consumers or professional (pro) users—Apple was to build computers for people, not corporations. Steve Jobs observed that: *"Our industry was in a coma. It reminded me of Detroit in the '70s, when American cars were boats on wheels."* He was convinced that Apple would be great again if it could merge the ease of use and elegance of the Macintosh with the power of the new Internet.

The launch of the *iMac* on May 6, 1998, a product kept so secret that few even inside Apple knew of it, was classic Jobsian showmanship. So too was the look of the computer: the monitor, all the circuits and even the modem were all in one box, harking back to the days of the early Macintosh. Like the Macintosh, the iMac was innovative in several ways. It was the first mass-produced computer to be *legacy-free*, an industry term denoting that whilst it may be able to connect with new peripherals like printers and cameras, it couldn't connect with older (legacy) devices. The iMac was the first computer to have the now-standard USB port and not to have a floppy disk drive at all. Apple has consistently been the first company to throw away old technology and embrace the new. The iMacs also featured bright candy-colored translucent plastic casings. These computers were definitely consumer electronics, not office products.

iMac's showing the color range

Forbes magazine described the candy-colored line of iMacs as an *"industry-altering success."* There were some other firsts for the iMac: it was the first Apple product to use the now familiar *i-product* designator, signifying *Internet*, and it was the first product that British industrial designer Jonathan Ive designed for Apple.

The public loved them—they looked cool, and unlike other PCs of the time they were easy to set up. An Apple TV commercial showed a seven-year-old kid and his dog setting up a new 1998 iMac in a time trial against a Stanford MBA student setting up a Windows PC. The kid had his iMac out the box, up and running and connected to the Internet in eight minutes, fifteen seconds. The graduate student took almost half an hour. Okay, the dog must have helped the kid, but you get the idea.[2]

Apple was back, not with just a product but with a philosophy that would guide it products to this day:

- Offer a small simple product range, where each product in the range has clear differences and benefits. Moreover, keep the range stable over time, so once customers have understood the product range they don't have to constantly relearn it.

- Focus on user experience; what people use the devices for rather than technical specifications. If you pay attention to Apple's advertising you'll see that the specs are rarely mentioned.

- Produce beautiful, simple designs that make people feel good owning and using the products. This applies to the hardware and the software.

- Insist on the highest build quality.

- Embrace innovation.

ALL YOUR MUSIC IN YOUR POCKET

As I've said before, you need a bit of luck to become a billionaire. An ex-employee of the electronics giant *Philips*, Tony Fadell, had an idea; he wanted to combine an MP3 player with a Napster-like music source. He hawked his idea around several companies, with no success, but it was just what Apple was looking for. Steve Jobs loved music; he used do date the folk singer Joan Baez, and Ella Fitzgerald sang at his thirtieth birthday. Apparently he had been shown *Creative's* new MP3 player the *Nomad Jukebox*, which by coincidence I used to own.

---

[2] You can watch this commercial called *Simplicity Shootout* on YouTube http://youtu.be/DmjvgOAhC_4

My Creative Nomad Jukebox MP3 player

Basically it looks like a Sony Discman, but contains a portable hard drive instead of a CD drive. Mine could hold six Gigabytes; that's about 80 albums of music. I loved it; this was a transformative device, just as my first digital camera had been in 1995. Imagine travelling with 80 CDs in your bag; the Nomad was smaller than one of the plastic wallets you might use just to hold that many CDs. If you set it to *shuffle* it was like having your own personal radio station that only ever played songs you liked with no irritating chat or adverts. You could create a playlist for a party and never have to touch it for hours, or even days if your parties lasted that long! Okay, it had its faults; the battery life sucked, and yes, its interface was hard to use.

Steve Jobs apparently really hated it. In particular he loathed the interface, but he also saw that the new 3.5 inch hard disk drives Apple had been looking at for its new laptops would be perfect in a portable MP3 player—half the size of the drive in the Nomad. Jobs realized that if Apple could get to market first with a cigarette packet-sized MP3 player, that was easy to use and that could connect to an online music store, like Fadell had envisaged, then Apple could dominate this market.

"Hold on", you're saying, *"why are we talking about MP3 players? They're not computers!"* Well, no they're not, but they do contain computer chips; they are an example of computers transforming the way we do things. For example, digital cameras have transformed the way we all take and share

photos—no more buying expensive rolls of film, taking a few precious photos, waiting for the roll to be full, sending the film off to be developed and finally getting 24 photos back, half of which are blurry and badly composed. Now we shoot dozens of photos, delete the ones we don't like, crop, enhance and improve the ones we do, and then share them with our friends on *Flickr* and *Facebook*. In a similar way, the iPod has transformed the way we listen to and buy music.

Jobs was very focused on ensuring the iPod was a success. Once again Jonathan Ive, now Vice President of Apple's *Industrial Design Group*, led the design of the iPod. Its interface, featuring a scroll wheel that allowed quick access to any song, was a significant innovation. It was reported, *"Steve would be horribly offended if he couldn't get to the song he wanted in less than three pushes of a button."* Released in time for Christmas 2001, the iPod was a big hit, even though it was only then compatible with Macs and the iTunes music store wouldn't open for another two years. It was, as Jobs had envisaged, smaller and much easier to use than the competition. Moby, the musician, noted, *"I've had three MP3 players, and I haven't figured out how to use any of them. And, this one, I held it and 45 seconds later, I knew how to use it."*

Just as Xerox could have owned the PC market, digital music is a business Sony should have owned. Sony invented portable music players with its Walkman (I owned several) and it continued to dominate the market even after dozens of other companies sold Walkman and Discman copies—the Sony devices were always the cool ones. In addition to being an electronics company Sony also owned several of the world's largest record labels, including: Columbia Records, RCA and Epic. But, in trying to protect its music business, Sony crippled its digital music players. Its digital Walkman couldn't even play MP3 files, even though that was the dominant standard for digital music. Although Apple prefers its technically better AAC format, all iPods support MP3 as well.

Jobs was finding dealing with the music industry as difficult as he had Hollywood; they were very suspicious of MP3 music since they'd taken down Napster. One by one, Jobs persuaded the big music companies to come over to iTunes, perhaps because they didn't think iTunes would sell many songs. In 2008 it became the number-one music seller in the US, and whereas those of us from an older generation first bought our music from a local record store, kids today will all buy their first songs from iTunes.

I remember around 2003 I took my Nomad (the one Jobs hated) into a hi-fi shop; I wanted to hear how it would sound when connected to a new amplifier and speakers I was thinking of buying. The sales assistant was very sniffy about me plugging in this little MP3 player rather than a high-fidelity CD player as a music source. Today most music systems have built-in iPod docks, and many people don't even own their music in a physical format. Jobs and Apple have transformed an industry.

By 2007 almost half of Apple's revenues were being generated by its pocket music player. Jobs' hunch had been correct —the iPod had 74% of the MP3 player market share and was the next big thing! The iPod also created a *halo effect* for Apple, encouraging Windows users to consider Macs, even though iTunes has long been compatible with Windows. The iPod encouraged people to go into Apple's bright, friendly stores, where they'd check out the new shiny Macs at the same time. This halo would be magnified when the iPhone was released, and then again with the iPad.

I WANTED ONE SO MUCH IT HURT?

It's January 9th 2007 and you are at the Macworld conference in San Francisco, waiting for Steve Jobs to give his much-anticipated keynote presentation. People camped out overnight to get a seat, but you've got a VIP pass, lucky you. The auditorium is packed, standing room only. The rumors are that Jobs it going to unveil a new Apple product—the smart money is on an Apple phone.

James Browne's *I Got You (I Feel Good)* is playing on the PA and as the song ends, Jobs enters the stage wearing his trademark black turtleneck sweater, faded blue Levis and white New Balance trainers. He looks happy. As the applause dies down he says, *"Thank you for coming, we're going to make some history here today."* After making some general statements about the Mac, he pokes fun at Microsoft and plays a new *I'm a Mac, I'm a PC* commercial about Windows Vista. This series of commercials has been hugely successful, showing an urbane, slim young man (the Mac) making fun of a chubby, bumbling, geeky man (the PC). Mac users love them, reinforcing the cool chic of owning a Mac.

Steve Jobs at Macworld San Francisco

Jobs then goes on to talk about the iPod range, emphasizing that the iPod is the world's best selling MP3 player. He then shows some stats showing that Apple has sold over two billion songs via its iTunes music store. Apple is now the fourth largest music retailer, and is expanding into selling TV shows and movies through iTunes. Jobs ends the segment with another dig at Microsoft, showing that its competing *Zune* music player has only 2% of market share.

Jobs is famous for his keynote speeches; he appears effortless, natural and unrehearsed despite the reality that these presentations are very carefully choreographed and involve a large team of people backstage. He gives a faultless demonstration of the new Apple TV system and then brings the tone down. Against a backdrop of the Apple logo eclipsing the sun, he says in a serious manner, "*This is a day I've been waiting for two and a half years. Every once in a while a revolutionary product comes along that changes everything.*" It seems that Jobs has turned on his famous *reality distortion field* to maximum power and is dragging everyone in the room into its influence. He makes the point that Apple has been very lucky to introduce several revolutionary products over the years. In 1984 it was the Macintosh, in 2001 it was the iPod, and today Apple is unveiling three revolutionary devices: a widescreen iPod, a revolutionary mobile phone and an Internet communication device. Jobs then plays with the audience, repeating: "*an iPod, a phone, and a internet communication device*" over and over until he says, "*do you get it? These aren't three separate devices. They are one and we are calling it the iPhone.*" The crowd goes nuts, cheering and applauding. The rumors circulating in the press were true; Apple is entering the fiercely competitive mobile phone market. "*Today Apple is going to reinvent the phone and here it is*" Jobs says, revealing a joke photo of an iPod with a circular dial like an old rotary telephone. He's playing the audience masterfully—everyone in the room is hanging on his every word.

Job's old friend Woz is sitting in the auditorium, along with the 1984 Macintosh team whom Jobs has invited to share in Apple's triumph. If you could see Woz, you would see that he is wiping tears from his eyes. He's happy for his friend Jobs and happy for the company that he co-founded and still loves.

The development process for the iPhone had started years before in 2005 when Jobs had directed Apple engineers to start looking at touchscreens. Everyone was envisaging a tablet computer, but once Jobs saw an early demo of the screen scrolling numbers, he realized it would be suitable for a remarkable new phone. He knew that if phone manufactures added MP3 player capability to their phones iPod sales would plummet. Nobody would need to carry a phone and an MP3 player. But, unlike other high-end phones, the Apple phone wouldn't let you browse the web poorly, take notes with a miniature keyboard and check your emails with difficulty. This would be a computer with a cell phone inside. This is what distinguished the iPhone from all other *smartphones* at the time. They were all cell phones that had some computing functionality. The iPhone was a pocket-sized computer; it even ran the Mac operating system OS X, though in a version customized for the touchscreen called iOS.

Not surprisingly, I've been a keen user of PDAs for years; owning first a *Psion* and then several *Palm Pilots* and even an *HP Pocket PC* for a while. I've also owned a digital camera since 1995 (in marketing jargon I'm an *early adopter*), an MP3 player, and of course I had a cell phone. So I commonly carried four digital devices with me. For years I had in my mind's eye the perfect digital device. It would combine the functionality of all four of the gadgets above in a single device. It would fit into my pocket, have a good battery life, a reasonable camera, enough on-board memory to store a 100 or so music albums plus a good number of photos, and I'd need to be able to access the Internet via it. For years, as my friends bought expensive Nokia's, Sony Eriksson's and Motorola's that combined some of my perfect device's functionality, but crucially not all of it, I held out and carried a separate PDA, MP3 player, camera and phone. That is until I saw the iPhone. Here was my perfect device in an unbelievably small, sexy package.

However, there was a big problem—the first generation iPhone wasn't for sale in New Zealand. It was never made available in New Zealand! I wanted one so much it hurt. I learned that if you had some technical skills you could *jailbreak* an iPhone; that is, release it from its factory tethering to AT&T's US cell phone network so it could work anywhere on the planet. I called a friend in New York, who went down to the Apple store in Manhattan, where she bought an iPhone and FedEx-ed it to me in New Zealand (thanks, Cat). A few days later I unboxed my new iPhone and set about jailbreaking it. Today, jailbreaking an iPhone is an easy task, with a software application that automates the process in minutes. Back then, the process took hours and required accurately typing instructions into the tiny iOS terminal window to reprogram the iPhone. One screw-up and you were left with an expensive paperweight. This disaster is called *bricking* your iPhone. I was very careful and was successful, and was rewarded with being the proud owner of one of the first iPhones in New Zealand.

Unboxing my iPhone, Sept 2007

Once again you're saying, "*hold on, what's with the iPhone? This book's about computers!*" Well, my point is that the iPhone marks a major point in the evolution of computers. Before the iPhone, computers were confined to offices and desks, and laptops with poor battery life. They mostly did tasks like writing documents, analyzing spreadsheets and databases, whilst at home people used them to browse the Web, check their email, manage their photos and music libraries, and to play games. But with the advent of the iPhone, powerful computers became very portable and importantly, they were constantly connected to the Internet. It's this combination of portable powerful com-puting power (the 1998 iMac had a 233 MHz computer chip whereas the 2010 iPhone 4 with a 800 MHz chip is almost four times as fast) and constant connection to the Internet that marks the beginning of what scientists call *ubiquitous* computing. This is where we are all constantly connected and have computing power at our beck and call everywhere, 24/7. This is changing our lives in ways we're often not aware of.

For example, earlier this year my wife and I went to a friend's wedding on the island of Rarotonga in the Cook Islands; about four hours flight north of New Zealand. Rarotonga is a beautiful tropical island: palm trees, sandy beaches, coral reefs, tropical fish - the whole package. There was a large group of us all attending the same wedding, and obviously we spent time together. It was remarkable how often in conversation one or more of us would want to check a fact: who directed that movie, what was the address of that charming restaurant, when was that show scheduled? Only to reach for our smartphones and then realize Rarotonga might be idyllic, but it doesn't have any Internet access for phones! It was so frustrating; it felt like your memory had been wiped. At home, we had all become so used to constantly using the Web as an enhancement to our memory and as a problem-solving device, that without it we all felt less capable. Some people already believe that this is the start of humanity become part biology and part machine—there will be more on this in the final chapters.

POST-PC

"If I were running Apple, I would milk the Macintosh for all it's worth – and get busy on the next great thing. The PC wars are over. Done. Microsoft won a long time ago."
—Steve Jobs before Apple bought NeXT

It's January 27 2010; Steve Jobs is once again on stage and there's a strong feeling of déjà vu. He's wearing his trademark jeans and black turtleneck, but this time he's launching the iPad. Apple still makes PCs; in fact its market share has been increasing despite the hard economic times, partly because of the halo effect from the iPod and iPhone, and because its new range of Macs are so stylish. Despite this, Apple is embracing the *post-PC* era. Sure, computers will still be needed in offices and businesses, but at home? Jobs doesn't think so; he believes that most computers are too difficult for the average person to use, and that they get in the way of the things people want to do. To this end Apple created a tablet computer, the iPad, which uses the same touchscreen interface as the iPhone. Despite the fact that many media commentators doubted the market for the device, jokingly calling it an *iPhone-mega* but with no phone, it went on to become the fastest selling consumer electronic device in history, selling over one million in the first 28 days. Fifteen million iPads went on to be sold in the first nine months making it the most successful consumer product launch in history. Apple's stock soared to over $400, making it the highest valued company in the world; valued at over $150 billion dollars, eclipsing Exon Mobil, and for a time Apple held more cash reserves than the US Government. Not bad for a company started in 1976 by two hippies in Silicon Valley.

Alan Kay with a mockup of the Dynabook

The iPad can trace its ancestry back slightly further. In 1968 Alan Kay, then working for SRI, described what he called the *Dynabook*. This was a small tablet or slate computer that looks very much like the iPad (except it has a physical keyboard below the screen). Its target audience was children since it would be so easy to use; children could interact with the Dynabook directly through its screen. Back in the early 1970s it wasn't possible to build a Dynabook. Now Apple has realized Kay's vision.

Things couldn't be going better for Jobs. He was financially secure, having made billions from Pixar and even more now from Apple. He was applauded as one of the greatest business leaders of any generation and as tech visionary. His personal life was also better than ever; he was happily married to Laurene, had three children and was even reconciled with his first daughter Lisa, who sometimes lived in their Palo Alto home. Life, though, has a funny habit of putting what's really important into focus. Jobs was fabulously wealthy: perhaps worth $10 billion; but without health, what use is money? In 2004 Jobs announced to Apple employees that he'd been diagnosed with pancreatic cancer. For the next five years, whilst he had some time off, he continued to work himself hard, and often appeared thin and gaunt in public. In 2009 it was announced he had received a liver transplant and although his prognosis was pronounced as "*excellent*", in 2011 at the launch of the iPad 2 he appeared thinner than ever.

Jobs' stubborn "*I know best*" attitude that had so often served him well in the tech world didn't help him in his fight against cancer. Walter Isaacson reports in Jobs' official biography "*Steve Jobs,*" that he initially refused an operation to remove the tumor and tried to heal himself by meditation, strange diets and other alternative treatments, against the advice of his physicians and his loved ones. The tumor spread to his liver, which necessitated the transplant in 2009. But this radical surgery was also too late and combined with his vegan diet, and immense workload his body was under severe pressure.

A few weeks before I wrote this, in August 2011, Jobs announced his resignation as CEO of Apple "*so he could focus on his health.*" Commentators wondered what would become of Apple without Jobs' guiding hand and vision.

Steve Jobs at the intersection of technology and liberal arts
(This photo was taken by Flickr user *backofthenapkin*)

"We've always tried to be at the intersection of technology and liberal arts, to be able to get the best of both, to make extremely advanced products from a technology point of view, but also have them be intuitive, easy to use, fun to use, so that they really fit the users – the users don't have to come to them, they come to the user."

—Steve Jobs

It's certainly true that he took his interests and personality traits (electronics and font design), obsessiveness and perfectionism, and turned them into a career. At Apple events there is often a signpost that marks a hypothetical street intersection between technology and liberal arts. Jobs passionately believed this is where a company must be located if it is to create great products. Creativity in both the arts and technology is about personal expression. Jobs was fond of quoting Henry Ford, who once said: *"If I'd asked my customers what they wanted, they'd have said a faster horse."*

Products in development at Apple aren't passed from the design team to the engineering team, to the programming team, and finally to the marketing team. Their design process isn't sequential. Instead, all groups work on a product concurrently and collaboratively. Jobs' pursuit of excellence in design is legendary; for him, design isn't decoration. It isn't the appearance of a product, it's the way the product works— function, not form.

Consider two examples. Have you ever visited an Apple store? If so, you'll have noticed a very striking glass staircase going up to the second and even third floors. Structural engineer James O'Callaghan designed this beautiful staircase, but its design patent has Jobs' name on it first. So, the staircase is pretty cool, so what? Did you walk up the staircase? It turns out that studies of retail stores show that almost two-thirds of shoppers never venture upstairs, except in Apple stores, where everyone wants to walk on the amazing glass staircase. The Manhattan Fifth Avenue Apple store, which opened in 2006 grosses more per square foot than any store in the world! Design as function, not form.

The power cord on a MacBook doesn't plug into a socket like on other manufacturers laptops; instead it attaches magnetically. A gimmick perhaps; until someone accidentally catches your power cord with their feet when walking by you, and your laptop isn't pulled onto the floor by them. You see, the magnet is designed to easily release, saving your computer from damage. Design as function, not form. Apple actually got the idea of the *MagSafe* magnetic power cord from a Japanese manufacturer of rice cookers who'd invented it to prevent children being burned by a cooker full of boiling water. As Jobs often said, quoting Picasso, *"Good artists copy; great artists steal."*

Innovation frequently originates outside Apple, and it's true that Jobs took existing technologies and made them easy to use. He took technologies out of the lab and put them in the hands of ordinary users. This process started with the graphical user interface, which Jobs first saw during his visit to Xerox PARC in 1979 that he turned into the Macintosh's GUI and has continued with Siri, Apple's new voice assistant for the iPhone, which Apple bought in 2010.

Jobs also understood that people rarely buy computers because of the hardware; they're interested what they can do with the machine. As we've seen before, a killer app is often enough to guarantee the success of the hardware it runs on. The Apple II was a hit thanks to VisiCalc, the first spreadsheet; the Macintosh took off because of desktop publishing, the iMac because of the Web, the iPod because of iTunes, the iPad because of the App store.

So if Jobs was a visionary, why did he spend time in the wilderness in the late 1980s and early 1990s? Perhaps because he was ahead of the times; his values of excellent design and usability were the wrong values when the early PC market was all about selling to corporations, which demanded low prices, not elegance and ease of use. The growth market now, though, is with home consumers and digital entertainment—the universal machine is no longer just a personal computer, it has entered our personal lives and is here to stay.

RIP

As I finished writing this chapter, Steve Jobs' death was announced. Apple.com's homepage showed an appropriately stylish photograph of Jobs looking intense and curious. The simple black and white photograph bore the simple caption, "*Steve Jobs 1955–2011.*" Clicking through the photograph brings you to a *remembrance* page where Apple's loyal customers are invited to leave their messages; over a million people have done so. One typical message reads:

"What Steve Jobs meant to me
This is equivalent to my mom's generation of Elvis dying for me. I am very sadden and emotionally moved at the moment. He was more influential on my life than my

parents and friends. While my parents loved me and friends shared fun times. Steve influenced me, motivated me to become the innovated, creative technologist I have become. I got into computer technology in 1980 and moved to Silicon Valley because of him. I have been one of his biggest admirers and looked to him as a mentor to push the boundaries of my own creative abilities to develop technology solutions which I hope made a difference and impact to the industries I worked in. We've lost a significant influence and icon in technology. We won't see another person of his innovation and foresight within my life time. He was the Edison of technology. He was and is one of my biggest inspirations. I feel I have lost a close family member."

Twitter and the blogosphere was alive with tributes, obituaries and comments; even the US President had something to say:

"Michelle and I are saddened to learn of the passing of Steve Jobs. Steve was among the greatest of American innovators - brave enough to think differently, bold enough to believe he could change the world, and talented enough to do it.

"By building one of the planet's most successful companies from his garage, he exemplified the spirit of American ingenuity. By making computers personal and putting the Internet in our pockets, he made the information revolution not only accessible, but also intuitive and fun. And by turning his talents to storytelling, he has brought joy to millions of children and grownups alike. Steve was fond of saying that he lived every day like it was his last. Because he did, he transformed our lives, redefined entire industries, and achieved one of the rarest feats in human history: he changed the way each of us sees the world.

"The world has lost a visionary. And there may be no greater tribute to Steve's success than the fact that much of the world learned of his passing on a device he invented. Michelle and I send our thoughts and prayers to Steve's wife Laurene, his family, and all those who loved him."

—President Obama

One thing is certain: Steve Jobs will not be forgotten—he changed the way we live with our universal machines. It's only fitting to leave the last word to him.

"The great thing is that Apple's DNA hasn't changed. The place where Apple has been standing for the last two decades is exactly where computer technology and the consumer electronics markets are converging. So it's not like we're having to cross the river to go somewhere else; the other side of the river is coming to us."

—Steve Jobs

Chapter 11

WEB 2.0

"When you give everyone a voice and give people power, the system usually ends up in a really good place. So, what we view our role as, is giving people that power"
—Mark Zuckerberg

I t's March 2011, and you are in the ancient city of Dara'a in southern Syria near the Jordanian border. This city has seen its fair share of turmoil through the ages. The ancient Egyptians conquered the region around 1450 BC as their empire spread across the Middle East. The Babylonians came, followed by Alexander the Great and the Greeks, who in their turn gave way to the Romans and then the Arabs. In the Middle Ages, Crusaders swept down from Europe before finally being beaten back to give way to the Ottoman Empire. Both of the twentieth Century's wars raged around the city and now in 2011, conflict stalks its ancient streets again.

Earlier in the month the security forces of Syrian President Bashar Al-Assad arrested 15 children for writing anti-government graffiti on their school wall. Some of the children were as young as nine, and it was rumored they were being tortured. On March 15, after days of negotiation by community elders failed to win the release of the children, hundreds of townsfolk gathered in front of a mosque to protest. News of the demonstration rapidly spread by mobile phone and SMS message; thousands joined the demo calling for reforms and an end to government corruption.

Suddenly shots rang out, the crowd panicked, many running for cover inside the mosque. When the crowd finally dispersed, four people had been killed. Their funerals that

I. Watson, *The Universal Machine,*
DOI 10.1007/978-3-642-28102-0_11,
© Springer-Verlag Berlin Heidelberg 2012

quickly followed developed into even bigger demonstrations, but the security forces were in an evil mood. Snipers were placed on rooftops and many more people were killed, but still the demonstrations grew and then spread to other cities across Syria.

The *Arab Spring* had started in Tunisia and spread to Egypt, Libya, Bahrain, Yemen and now Syria. The corrupt Tunisian and Egyptian governments had already fallen. Colonel Gadhafi had been shot in the head after a bloody civil war. The Assad regime is not going to hand over power lightly. Foreign journalists have been expelled from Syria. The regime shuts down the telephone systems and Internet whenever it chooses. Soldiers have instructions to shoot to kill anyone seen with a camera. Yet still video is appearing daily on YouTube and is publicized via Facebook and Twitter.

You learn from a young Syrian man that his uncle, who lives in London, has sent him a satellite phone. Explorers and adventurers normally use these phones whilst they are in the wilderness, or wealthy bankers on their yachts. However, they are just as useful inside a repressive state that is trying to block communication with the outside world. Syrian ex-patriots are sending these phones, costing around $1,000, into Syria so the world can witness what Assad's regime is doing to its own people. Cell phone cameras can be easily hidden inside clothing and vehicles, and are used to secretly record as security forces beat and shoot innocent civilians. They even record tanks prowling the streets shelling mosques where the injured are being treated. The illicit video can then be transferred to a laptop, and sent via satellite phone to YouTube, where Facebook and Twitter can spread the truth to the Syrian rebels and the entire world. The *Guardian* news-paper reports that, "*upwards of 30,000 videos have now been posted by Syrian opposition activists.*"[1]

The way this remarkable social infrastructure came about is the subject of this chapter.

---

[1] http://www.guardian.co.uk/world/2011/dec/13/syria-torture-evidence

LET ME SHARE WITH YOU

The web has always been about publishing and sharing information; that after all is why Tim Berners-Lee invented it. In the beginning you had to have access to a web server in order to publish a website—that's why most early users were computer science students. However, very soon companies were established that allowed anyone to easily create and manage their own websites. *GeoCities*, established in 1995, was one such small web hosting company. It allowed users to establish websites initially within six themed neighborhoods, including: *Coliseum*, *Hollywood*, *RiverdaleDrive*, *SunsetStrip*, *WallStreet*, and *WestHollywood*. Users would choose a neighborhood within which to create their website, which became part of their unique web address. Bulletin boards, chat and other community services were quickly added, and the number of users grew rapidly. By June 1997 GeoCities was the fifth most popular website.

In 1999, at the height of the dot-com bubble, *Yahoo!* purchased GeoCities for $3.5 billion. Yahoo! badly mismanaged its new acquisition, changing the terms of service for the website, claiming it owned all the users' content— even personal photographs they'd uploaded. This was very unpopular, and users started to leave in droves. In 2001 false rumors circulated that Yahoo! was going to close GeoCities. Over the following years users continued to drift away to competitors' websites, and in 2009 Yahoo! shut it down. Rupert Goodwins, a journalist, described GeoCities as: "*the first proof that you could have something really popular and still not make any money on the Internet*".

As GeoCities was being mismanaged by Yahoo! the competition picked up the slack, in particular sites aimed at young people. Two computer programmers started *Friendster* in 2002. The name was created by combining "*friend*" and "*Napster;*" it aimed to provide a safe, legal place for people to meet and socialize. Users created profiles and linked to their friends' profiles, and friends-of-friends, and so on. The site quickly attracted several million users, and following good publicity in magazines like *Esquire* and *Time* its user base continued to grow, peaking at over 100 million in 2008.

Others saw Friendster's rapid success and emulated its model. One well-resourced clone was *MySpace*, which launched a year later in 2003. MySpace was infinitely customizable, allowing people to express themselves though the design and content of their MySpace page. MySpace was particularly focused on music, which its owners believed would attract a youthful demographic. It provided tools to upload music to the site, and allowed users to share and rank the music. *News Corp* bought MySpace for almost $600 million in 2005. The dotcom bubble had burst, and this was reflected in its bargain price in comparison to Friendster a few years earlier. By 2006, most musicians and bands had a MySpace fan page and the site had over 100 million users. Social networking was becoming the engine of growth for the Web.

By 2008 there were two big social networks: Friendster and MySpace, each with over 100 million users. There were also other sites, like Google's *Orkut,* which was big in Brazil but nowhere else, and others like *Bebo*, popular with younger kids, *Hi5*, *Ning*, *Plaxo*, *SixApart*, and more, but you may never heard of any of them. The social networking site you've certainly heard of, and probably even use, is *Facebook*. The story of how it was created has become legendary.

HARVARD'S FACE BOOKS

Harvard University can boast at least two alumni who have founded enormously successful computer businesses without actually ever graduating. The first is *Bill Gates*, whom you've heard of, and the second is *Mark Zuckerberg*, whom you may not know. Zuckerberg was born in 1984 in White Plains, New York. He is really the first character in this book who is a child of the PC era—the Macintosh was released the year he was born. His parents are medical professionals and he went to the exclusive *Phillips Exeter Academy,* where he excelled in science and classics. He'd had a computer since he was a child, and was a talented programmer and liked to make his own computer games. At Exeter he even took a graduate course in computing from a local college. Whilst still at school he created a music player, called *Synapse that* learnt users' musical preferences using artificial intelligence—it got favorable reviews in several computer magazines. *AOL* and *Microsoft* saw the reviews, and both tried to buy Synapse and to hire Zuckerberg—it's rumored for over a million dollars! He declined their offers and enrolled at Harvard in 2002 as a computer science major.

Mark Zuckerberg (Photo by Guillaume Paumier)

At Harvard, in addition to his studies, Zuckerberg quickly set about creating interactive websites for other Harvard students to use. The first was *CourseMatch,* which enabled a student to select a class to study based on the class selection of other students. The official motivation for CourseMatch was to help students form study groups, but students really used the service to select courses that people they wanted to date were taking. CourseMatch was innocent enough, and

Zuckerberg gained a reputation at Harvard for being a gifted programmer, though somewhat nerdy and socially inept. His next project would make him infamous.

Harvard students live in dorms, and every dorm has a *Face Book* that contains the name and photograph of everyone in the dorm. The Face Books were available on the dorms' websites so students could access them from their computers. Each dorm's Face Book was private to members of that dorm. One night, whilst drinking beer and publicly recording his actions in his blog, Zuckerberg hacked into every Harvard dorm's website and illegally obtained their Face Books. He then created *Facemash*, a website that took a random pair of photos of two female students and asked the user to decide which of the pair was "*hotter.*" As users rated more and more random pairs of students' photos, an algorithm worked out a global ranking of all the photos and the *hottest* students could be identified. Zuckerberg released the website's URL to a few friends that weekend—Facemash instantly went viral at Harvard. By Monday, Zuckerberg was forced to switch it off and he was in BIG trouble.

Feminist groups on campus were outraged. The administrators of the dorms were shocked that the photos had been taken illegally. Facemash's popularity had overwhelmed Harvard's network, preventing staff and students accessing the Internet. Individual students were complaining their photos had been used without their permission. And, I'd guess more than a few students just didn't like their ranking! Zuckerberg was hauled in front of a disciplinary tribunal, who accepted that he'd acted without malice, and that Facemash could perhaps be attributed to his genius, naivety and social inexperience. He was put on probation— Zuckerberg was now notorious at Harvard.

IDENTICAL TWINS

Cameron and Tyler Winklevoss moved in completely different social circles to Zuckerberg. They were athletes, 6' 5" tall; Cameron was left-handed and his identical mirror-image twin brother was right-handed. Their powerful physiques and unique mirror-imageness made them ideal rowers; they were to go on to compete in the Olympics in the men's coxless pairs. From a wealthy family, intelligent, athletic and handsome, they moved amongst Harvard's elite as members of the ultra-exclusive men's-only *Porcellian Club*, which numbered a US President amongst its old boys.

Like many high achievers at the time the Winklevii, as some people called them (though not to their faces), along with a classmate Divya Narenda, wanted to start a website and make themselves even richer. They conceived a social networking site for Harvard students, which they planned to call *HarvardConnection*. Essentially it would be a dating site, which they would to roll out to other Ivy League universities if popular. The problem was, none of them had the technical skills to build the site, and the student they had hired to write the code had quit on them without finishing. Zuckerberg had the skills though; the twins contacted him to see if he'd be interested in building HarvardConnection for them.

Now here the story gets a bit complicated. Zuckerberg's version of events is that he met and spoke with the Winklevoss twins, saw the code they had already developed, but thought the functionality of HarvardConnection was too limited and left it there. The Winklevosses' version of events is quite different. They claim that Zuckerberg liked the idea of HarvardConnection, stole the code they had already written, delayed and prevaricated for a couple of months letting them believe he was working for them. But, all the time he was secretly developed his own version of HarvardConnection, which he released calling it *TheFacebook.com*. Isn't it amusing that identical twins think their software and Facebook are identical twins!

Zuckerberg claims he was inspired by an article in *The Harvard Crimson,* commenting on the Facemash debacle, which observed that there clearly was an appetite for a social networking tool for students. The December editorial commented, *"Put Online a Happy Face: Electronic facebook for the entire College should be both helpful and entertaining for all,"* and it practically went on to describe the feature set Harvard's facebook should have. Zuckerberg obliged by building one and on February 4, 2004, Zuckerberg with $1,000 funding from classmate Eduardo Saverin, launched thefacebook.com. Initially you had to have an *@harvard.edu* email address to join, and within a month more than 85% of Harvard undergraduates were signed up.

Facebook, as students just called it, was simple. You posted a photo on your profile, wrote a bit about your background, home town, schooling, likes and dislikes, and importantly, set your *"Relationship Status"* and then invited friends to connect to, or *friend* you. Students loved it, and recall checking in to Facebook first thing in the morning and last thing at night, to keep up with their friends' updates. Despite being called social networks, Friendster and MySpace were mostly about searching through lists of people you didn't know and trying to hook up with them. Facebook's point of difference was that you knew the people you *friended*. You might not know them well, but they would be at least friends-of-friends; you had someone in common. This was, and remains, Facebook's genius.

Realizing that they needed more people if Facebook were to expand to other colleges, Zuckerberg and Saverin recruited three classmates, Dustin Moskovits, Andrew McCollum and Chris Hughes, to help. Facebook was then rolled out to other Ivy League universities and Stanford on the west coast. Seeing that Facebook was fast becoming a success, the Winklevoss twins had a lawyer, who worked for their father's Wall St. consulting firm, write Zuckerberg a *"cease and desist"* letter, threatening legal action if he didn't take thefacebook. com off line. The Winklevii also pulled strings to get a meeting with the Harvard President, where they claimed that Zuckerberg should be penalized under Harvard's honor code that forbade students stealing from one another. The Harvard President wasn't interested in their plea, and dismissed them.

In June, just four months after Facebook's launch, Zuckerberg met with venture capitalists in New York and one offered him $10 million for the company—Zuckerberg declined.

THE NEXT BIG THING

Sean Parker had been busy since Napster had been bankrupted by the music industry. He'd founded *Plaxo,* an online address book synchronization service that had a social aspect to it, but had fallen out with the venture capitalists that backed it, and quit. Nonetheless, he had gained a reputation as being involved with two successful dotcom start-ups. Not many people in 2004 had that on their résumé! He was looking for the *next big thing*, and he knew it would involve social networking.

He'd checked out MySpace, but it was all about ego and narcissism: look at me! Look at my band, look at my videos, my photos, aren't I wonderful, pay attention to me! It really didn't have a strong social element; you didn't use MySpace for a conversation. Friendster on the other hand, under the covers, was just a dating site. Boys trawled through Friendster trying to get pretty girls' email addresses from them.

Sean Parker

In the 2011 triple Oscar-winning movie *The Social Network*, which is about the founding of Facebook, Justin Timberlake plays Sean Parker. In the movie, one morning Parker wakes in the bed of a beautiful young Stanford student. He goes to check his email on her computer and sees her Facebook page is open. He realizes that this is what he's been looking for. Facebook was so simple; all the people she was connected with actually *were* her friends, not randoms just trying to date her. Even while her computer was off, updates from friends would be posted, giving her a reason to constantly check back in to keep up with what was happening in her social network. It was genius, and across the bottom of every page was written: *"A Mark Zuckerberg production."* Parker had to meet him and get involved.

Parker found Zuckerberg through a Google search, and he arranged to meet with him and the rest of the Facebook team in a trendy Manhattan restaurant. Parker played the successful dotcom entrepreneur very well, and Zuckerberg liked him. Parker insisted that if Facebook wanted to be the next really big thing, they had to move out to Silicon Valley because that was where the dotcom scene was happening; where the talented programmers and venture capitalists where. Saverin wasn't convinced; he thought Parker was trying to muscle in on Facebook. Anyway, he planned to spend the summer holidays working as a Wall St. intern and raising advertising money for Facebook in New York.

Zuckerberg agreed with Parker and despite their disagreement, Saverin gave Facebook $18,000 so they could hire two interns (computer science students from Harvard) and rent a house on La Jennifer Way in Palo Alto for the summer. Saverin would stay in New York and raise funding. Facebook was formally incorporated in June 2004. Zuckerberg owned 51% of the stock, Saverin 33%, and Parker was made president with about a 7% share. Moskovits, McCollum and Hughes held the remaining stock.

Tales of wild parties at the Facebook house in Palo Alto that summer are legendary, and probably somewhat exaggerated. Whatever else the young men were doing, they were doing plenty of coding, as Facebook enhanced its functionality and expanded to colleges across the US and into Europe. Saverin failed to obtain funding in New York, but Parker, who had moved into the Facebook house, was instrumental in obtaining $500,000 from Peter Theil, who a few years earlier had made a fortune selling *PayPal*, which he co-founded, to *eBay*. The new funding diluted Saverin's stock, making it virtually worthless, and after angry scenes he left Facebook and sued.

The lawsuit was settled out of court for an undisclosed sum protected by a non-disclosure agreement; Saverin is still listed as a co-founder of Facebook. Sean Parker also didn't last long with Facebook. A year later police raided the house he had rented and was partying in, and he was arrested for cocaine possession. Although the charges were subsequently dropped, the investors in Facebook forced him out. Parker is now on the board of *Spotify*, a social music streaming service, but his legacy remains. In particular, when the new corporate structure of Facebook was established, Parker still smarting from his eviction from Plaxo, ensured the same could never happen to Zuckerberg at Facebook.

The Winklevoss twins had also filed a lawsuit against Facebook, claiming that Zuckerberg had broken a verbal contract with them, that he had copied their ideas and had stolen their code. Four years later, by which time Facebook had 100 million users, they settled reportedly for $65 million.

At the time of writing (December 2011) Facebook has over 800 million users. It seems every one has a Facebook page, corporations, this book,[2] even the British Queen has a Facebook page.[3] MySpace's user-base has crashed and News Corp is allegedly looking for a buyer to cut its losses; Friendster has also lost most of its users. Unusually for a dotcom startup Facebook is still privately owned; Zuckerberg has turned down repeated offers to buy it and it's now valued by some at $100 billion, making Zuckerberg the world's youngest billionaire. However, like many of the fabulously wealthy inventors in this book, he isn't motivated by money and

---

[2] http://www.facebook.com/UniversalMachine
[3] www.facebook.com/TheBritishMonarchy

doesn't seem at all interested in it. He's been voted Silicon Valley's worst dresser by *GQ* magazine; he doesn't own a fleet of expensive sports cars, yachts, a Caribbean island, or date supermodels and movie stars. In fact he lives modestly in a rented house, and has joined Bill Gates in pledging to give at least half of his wealth to charity. He encourages other wealthy people to join him. Sean Parker is also know for his philanthropy and is the founder of *Causes,*[4] which uses social networking to link charities with potential donors and then broadcasts the connection to their social network—it raises over $20,000 a day for a wide range of charities.

Many people, myself included, don't use Facebook in their professional lives, reserving it strictly for their social connections. LinkedIn, launched in 2003 by Reid Hoffman and other ex-PayPal employees, now reports more than 120 million users; it is the dominant business social networking site and the one I use.[5] LinkedIn uses the power of social networks to foster business opportunities. You can reconnect with old work colleagues, look for a new job or seek advice and assistance with current projects. I even found the publisher for this book through LinkedIn.

I joined the LinkedIn group *Popular Science Book Professionals* and posted a description of this book and asked if anyone could, "*recommend an agent or publisher.*" Angela Lahee, an editor at Springer in Germany, replied saying she'd seen my book proposal and my blog and was interested. So through the power of social networking I got a publishing contract. I asked Angela to comment for this chapter and she said, "*as a science publisher one not only has to keep up to date on important new developments in research, but also – and particularly in popular science publishing – to be constantly alert to unexpected opportunities. This latter was a part of my motivation for joining the LinkedIn group. Of course I was very happy when you responded positively to my enquiry.*"

---

[4] http://www.causes.com
[5] My public profile on LinkedIn is http://nz.linkedin.com/in/drianwatson

AMBIENT INTIMACY

Privacy has become a *cause célèbre* for social networks. When Google launched its *Buzz* social network, it created a storm of protest when everyone who had a Gmail account woke up one morning to find all of the people in their address book were now automatically following them in Buzz. Your boss knew your location; your ex-wife could follow any comments you made online. Google rapidly backtracked and made the service *opt-in* instead of *opt-out*, but Buzz never really recovered from its calamitous start. A year later Google launched *Google+,* a new social network, much more carefully to a warmer reception—Google has learnt to treat privacy very seriously. [6]

Facebook itself came under repeated criticism for having privacy options that were too complicated, defaulted everything to public and were a pain to switch off. Of course you could argue that any provider of a social network clearly has a conflict of interest; they want the network to default to being public, since if everything were kept private, then there would be no social aspect to the network.

I take a slightly different view of privacy to many people. You see, I believe the concept of *privacy* is a somewhat modern notion. If you've ever lived in a small village you'll know what I mean. When I lived in England, for many years I lived in a small village northwest of Manchester. The population was about 300 and we boasted four pubs (well actually, one was technically in the neighboring village but was closer to our village center than theirs). Conversation in the pubs amongst the locals revolved around sports, politics, the weather (always of interest in England), but often other peoples' *doings*, their comings and goings; their private business in other words. I quickly learned that you couldn't do anything without it almost instantly being known by everyone in the village who cared to find out. The concept of privacy and anonymity is, you see, a very modern idea that really only came about when people started to live in

---

[6] This book's Google + page is https://plus.google.com/u/0/b/117947364453412518342/

large cities, where they didn't know any of their neighbors. However, in small communities there never was privacy or anonymity, as everyone knew everyone else, their families, their history and their business.

Facebook has one crucial and significant difference to many Internet services in that your Facebook profile is based on your real identity. You cannot be anonymous on Facebook; if you create a false Facebook persona your real friends will not befriend you. So just as in a small village if you say something about somebody on Facebook they can see who said it; this is a very different ethos to the hacker cult of anonymity. Zuckerberg is very insistent that *"you only have one identity,"* and that you shouldn't maintain different personas for your friends, your family and co-workers. He believes that, *"having two identities for yourself is an example of a lack of integrity."* He also believes that as information about you permeates across the Internet it will increasingly be impossible for you to maintain distinct personal and professional profiles. This might explain why the young, who have grown up with Facebook, are not as worried about online privacy as their parents. They're quite happy setting their privacy to share with *"Friends-of-Friends"*, which if you think about is much the same as *"Everybody."* Zuckerberg's advice is, *"Privacy has not disappeared, but become even easier to control – what I want to share, I can share with everyone. What I want to keep private stays in my head."*

Steve Jobs was a friend and mentor to Zuckerberg, and like Jobs he often believes that he knows what customers want better than they do. Facebook has made mistakes in the way it has rolled out new features and managed privacy. Zuckerberg to his credit is not above personally apologizing, which he does on his own Facebook page.[7] In November 2011 he posted, *"I founded Facebook on the idea that people want to share and connect with people in their lives, but to do this everyone needs complete control over who they share with at all times...Overall, I think we have a good history of*

---

[7] Mark Zuckerberg's Facebook page is http://www.facebook.com/zuck

providing transparency and control over who can see your information. That said, I'm the first to admit that we've made a bunch of mistakes."

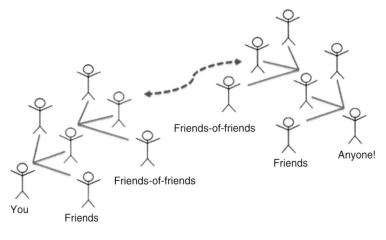

Four degrees of separation

Despite sometimes offending users Facebook continues to grow; a 2011 study by The University of Milan and Facebook[8] has shown that the old adage *"six degrees of separation"* that connects any two people on the planet is no longer true. The new study of 69 billion Facebook friendships (the largest social network study ever) shows that 92% of us are connected by only four degrees (five hops). Take a moment and think about this ... choose anybody, Barack Obama perhaps; it is probable that a friend-of-a-friend of yours knows a friend-of-a-friend of Obama's! The researcher also shows that within the same country most pairs of people are separated by only three degrees (four hops). Leisa Reichart, a technology commentator, has coined the phrase *"ambient intimacy"*[9] to describe how Facebook, Twitter and other social networking services allow us to remain in touch with friends, followers and those

---

[8] https://www.facebook.com/notes/facebook-data-team/anatomy-of-facebook/10150388519243859
[9] http://www.disambiguity.com/ambient-intimacy/

who share our interests in a way that was impossible just a decade ago.

LOLCATS

A popular feature of Facebook is the ability to easily upload and share photos with one's friends. Many dedicated online photo-sharing sites also exist: *Flickr*, which was bought by Yahoo!, and *Picasa*, bought by Google in 2004, are two popular examples. As the bandwidth available to everyone has continually improved so people have uploaded more and more photos. Instead of just uploading serious photos, some people uploaded frivolous and humorous photos and sometimes, through the power of social networking, some genres take on a life of their own.

LOLcats is a great example, featuring photos of cute or ugly cats, sometimes in unusual situations with humorous comments added to the photos. "*LOL*" is an abbreviation used in Internet chats and SMS-messaging for "*Laughing out Loud*," so a LOLcat is something funny about cats. LOLcats went viral in 2007 with a weblog (or blog) called *I Can Has Cheezburger* that featured LOLcats. The funny cats also started to show up on a video-sharing website called *YouTube*, which has become so indispensable one wonders how we ever lived without it.

An example of a LOLcat

Three ex-employees of PayPal, including one who had worked briefly at Facebook, founded YouTube in February 2005. YouTube's creation myth is that its founders wanted to share their own videos and were finding it hard, so they built a website to make the process easier. Within a year 65,000 videos were being uploaded daily. In 2007 people started uploading LOLcat videos of their cats being entertaining and the concept of the viral video was born. It's been estimated that in 2007 YouTube alone was responsible for as much band-width as the entire Internet in 2000. However, even though Google has bought YouTube for $1.65 billion it's not clear if it has ever made any money.

The vast majority of videos uploaded to YouTube are from private individuals, though in recent years large media corporations like the BBC, CBS and Hula have made some of their content available via YouTube. YouTube serves several functions in addition to letting us laugh at cats. It provides a way for us to share video from a family event, a child's first steps, graduation or wedding videos. It also acts as a collective social memory. People upload, illegally, video clips taken from their favorite TV Shows, movies, music videos and even TV commercials. There is hardly a great scene from a classic TV show, even ones dating back as far as the 1950s that is not available to watch on YouTube. Although the copyright holders complain, and YouTube does take down video that infringes someone's copyright, they just get uploaded again. YouTube forms a wonderful and unique cultural memory, allowing us to instantly, retrieve and watch everything from sublime comedy to great sporting moments.

YouTube of course has another, more serious use. I can't know what is making the news as you read this. But if you think of the name of a place where people are being oppressed and fighting for their rights, and enter that name into YouTube, you'll certainly find videos that protestors and activists have just uploaded, sometimes at considerable personal risk to themselves. Dictators and despots can no longer hide behind iron curtains or propaganda of their own making—the entire world is now their witness.

CITIZEN JOURNALISM

The Web has always been about publishing information; that was its original purpose. Maintaining a moderately well designed website, though, has always been a bit of pain, particularly if all you want is somewhere that you can keep a record of your thoughts and opinions, like a journal which you might write in daily. A basic issue is that we write from the top of the page to the bottom, and a webpage loads from top to bottom as well. This will mean that the latest content in a web journal will tend to be at the bottom of the webpage. However, a simple change let *web logs* (the origin of the word *blog*) appear in reverse chronological order, meaning the latest entry is always at the top of the page. There were several early *webloggers* in the mid 1990s, but Peter Merholz is credited with first using the diminutive form *blog* as both a noun and a verb in his blog Peterme.com in 1999.

Several blogging tools appeared almost simultaneously including *Open Diary*, *LiveJournal* and blogger.com, which was purchased by Google and is the blogging site I use for this book.[10] Blogging has taken off, and in 2011 it was estimated there were over 156 million blogs on the Web. Many of these blogs are just like someone's diary or journal; personal recollections of the day or the week, and are of little interest to anyone other than the author. Many are about very specific subjects: bird watching, quilting, you name it and there will be people blogging about it. A smaller number are written by specialists: journalists, economists, scientists, historians, financiers, politicians, and aim to provide authoritative comment and opinion. Some bloggers of course are complete crackpots, barking mad, but the Internet is as open to them as anyone.

Here again is the strength, power and weakness of the Web. Anyone can blog about anything. There is no editorial supervision or control. It's good that Syrians can publish their accounts of daily events inside Syria for the outside world to read. The Syrian regime can't stop them, though it would like to. But, and this is the other side of the coin, how we do we know that what is published in a blog is true and accurate? It could be false propaganda. If the *New York Times* publishes an article, I can be reasonably confident that the journalist checked the facts. The story has probably been passed through a chain of scrutiny and sign-offs before it ever gets published. Everyone involved in its publication has probably been to journalism school, and understands professional ethics. Ultimately, their jobs could be on the line if the piece turns out to false or defamatory. Citizen journalists, as

---

[10] http://universal-machine.blogspot.com/

bloggers are sometimes called, do not have to adhere to any such standards and face few sanctions if they do not.

So we have a dilemma, on the one hand we like blogs because of their openness and honesty; honest in the sense that they are directly of the people. But on the other hand, we should be always be on our guard and slightly suspicious, and ask ourselves what the blogger's motives are? There is also another issue, which Eli Pariser, a political activist in the US, calls the *filter bubble*. He noticed that different users get different responses from Google based on the history of their past Internet searches. This might not seem much of a problem, but consider this; suppose you had some crazy belief, let's say you firmly believe that the moon is made of cheese, and that scientists who say otherwise are lying to everyone in some grand conspiracy.

Before the Web you'd have been quite lonely; nobody would have shared your belief, and you'd have struggled to find any information to support your views. With the advent of the Web you can easily search for content that supports your belief; you may find blogs and forums and groups you can join full of people who also think the moon is made of cheese. Because of the filter bubble you'll not see articles that put forward an opposing viewpoint. You are happily living in a bubble of information that supports your crazy notion. Sincerely believing that the moon is made of cheese is harmless enough, but supposing a person believed the US government was planning to enslave the population, and that only a violent act against the government would cause the populace to rise up and secure their liberty. The filter bubble is now much more dangerous. People with extreme beliefs can now easily connect with like-minded individuals through social networking and blogs, and believe that their ideas are mainstream, normal even—this can be very dangerous.

140 CHARACTERS OR LESS

On the April 1st 2009 the British newspaper the *Guardian* announced that after 188 years of appearing in print, it was going to become the first newspaper in the world to publish exclusively on *Twitter*. All of the Guardian's news stories would be limited to just 140 characters, the maximum size of a Twitter message or *tweet*. They would also rewrite the paper's entire archive going back to 1821 as tweets. Stories they had already completed included: "*OMG Hitler invades Poland, allies declare war see tinyurl.com/b5x6e for more*" and, "*JFK assassin8d @ Dallas, def. heard 2nd gunshot from grassy knoll WTF?*" This was, of course, an April Fool's joke.

Twitter originated from a brainstorming session in the podcasting company *Odeo*. The idea was to take the way someone might use SMS to communicate with a group of friends and move it onto the Internet. The first version was launched within Odeo and rolled out to the public in July 2006. Tweets are a maximum of 140 characters long because they may be delivered by SMS in some countries as well as via the Internet. Tweets are publicly broadcast by default, but can be restricted to just a person's *followers*; people who have signed up to receive their tweets.

Twitter remained largely under the radar until the 2007 *South by Southwest* (SXSW) festival in Austin Texas, where Twitter had set up large video screens around the festival venues constantly streaming tweets. Festivalgoers noticed and started using Twitter at the festival to keep informed of what was happening. They then spread it to the rest of the world when they went home. 400,000 tweets were sent in 2007 but 400 million were sent in 2008. Twitter is a form of *micro-blogging* service, since it allows people to broadcast very short messages such as status updates. It is a very powerful tool for mobilizing people around an issue and informing a global audience of events and services. This book has a twitter account that keeps people informed on topics of relevance to the book.[11] Twitter now ranks amongst the top ten most visited websites, and like Facebook, Twitter has been very influential during protests in authoritarian regimes. The civil unrest in Moldavia in 2009, the protests around the 2010 Iranian elections, and the Arab Spring uprisings in 2011 for example, are sometimes referred to as the *Twitter Revolutions*, which brings us back to this chapter's opening.

As I was writing this, the media was full of speculation as to who would be awarded the 2011 *Nobel Peace Prize*. One of the names that cropped up quite often was Mark Zuckerberg, with journalists saying that the Arab Spring was only possible because of the way online social networks helped protestors to connect and broadcast. In the end the prize was shared between three champions of women's rights in Africa and the Middle East, but had Zuckerberg been awarded the prize, it would have made some sense.

My basic point is that computers have become more than tools that we use to perform specific tasks, like writing, editing photos, or researching something on the Web; they are now a conduit through which we live our lives. Web 1.0 was mostly a centralized broadcast medium; someone publicized information on a website and numerous people consumed that information via their browser. Web 1.0 was really just an online version of traditional print media. Web 2.0 is all about participation, user-generated content, sharing, collaboration, and linking virtual communities. The term "*Web 2.0*" was first used in 1999 by Darcy DiNucci in an article called *Fragmented Future*.

---

[11] @UniversMachine is this book's Twitter account, which you can follow at https://twitter.com/#!/UniversMachine

ALL THE WORLD'S KNOWLEDGE

Nothing perhaps encapsulates the philosophy of Web 2.0 as well as *Wikipedia*.

> *"Wikipedia is a free, web-based, collaborative, multilingual encyclopedia project supported by the non-profit Wikimedia Foundation. Its 19.8 million articles (over 3.7 million in English) have been written collaboratively by volunteers around the world. Almost all of its articles can be edited by anyone with access to the site, and it has about 90,000 regularly active contributors. As of July 2011, there are editions of Wikipedia in 282 languages. It has become the largest and most popular general reference work on the Internet, ranking seventh globally among all websites on Alexa and having 365 million readers. It is estimated that Wikipedia receives 2.7 billion monthly pageviews from the United States alone."*
>
> —www.wikipedia.org

Before Wikipedia the most famous encyclopedia was the *Encyclopaedia Britannica*, first published between 1768 and 1771 in Edinburgh. The print edition runs to 30 volumes, containing about 40 million words on half a million topics maintained by 100 full-time editors and 4,000 topic experts. Clearly this is a massive undertaking. It's now available on the Web and it's a great example of old school Web 1.0 thinking; there is a clear separation between information providers and information consumers.

Launched in 2001 by Jimmy Wales, an Internet entrepreneur, and Larry Sanger a philosopher, Wikipedia is the epitome of the ethos of the Internet and Web 2.0 in that it spreads information openly, collaboratively and is not for profit. Any user may edit any article anonymously; nobody owns an article and no article is vetted by any expert or authority figure. The content of an article is agreed by consensus and will change as each interested person edits it. Conventional wisdom would tell us that creating an encyclopedia in this way would cause chaos—surely somebody must be in charge? But apart from a handful of entries that are prone to vandalism, like President George W. Bush's entry, the system works just fine. Sure, Wikipedia suffers from the usual problems of user-generated content; namely that you can't be 100% sure an article is error-free. But ask yourself this: would you prefer an encyclopedia with 20 million entries, some of which may contain errors, or the *Encyclopaedia Britannica* with its half million entries, some of which also have been found to contain errors? Amusingly, Wikipedia has an entry called "*Errors in the Encyclopædia Britannica that have been corrected in Wikipedia*" that lists some notable "*mistakes and omissions.*" The Wikipedia entry says:

> "*These examples can serve as useful reminders of the fact that no encyclopedia can ever expect to be perfectly error-free (which is sometimes forgotten, especially when Wikipedia is compared to traditional encyclopedias), and as an illustration of the advantages of an editorial process where anybody can correct an error at any time.*"
>
> —www.wikipedia.org

The universal machine has now woven through how we interact with each other individually and collectively, at work and socially, as friends, colleagues and activists, as consumers of information and content creators. Our world is changing before our eyes.

Chapter 12

DIGITAL UNDERWORLD

*I*t's 2010, your name is Farzeen, you're an Iranian engineer working at the top secret and controversial Natanz nuclear research facility in Iran. You never feel relaxed when you're at work because you are worried that today might be when *The Great Satan,* or its Zionist puppet Israel, decides to attack the facility. You don't even feel safe at home now since an Iranian nuclear scientist was assassinated in Tehran, presumably by *Mossad*, Israel's spy agency. You briefly think about your student days at Leeds University in England—those were happy times. Iranians were popular on campus because the Shah had just been overthrown in a peoples' revolution. There was a good group of fellow Iranians at the University and so many pretty girls—there were some great parties.

You've just received a gift from a trusted colleague, a small USB memory stick, filled with some scientific papers you're interested in. You start up your laptop and plug the drive in. It contains the conference proceedings from a dozen recent international engineering conferences. There are hundreds of scientific papers; this is a great resource. It's very hard for you to travel to conferences any more, there's no foreign currency to pay for them and the security clearances are so hard to get. You can't view the papers online either because you have no Internet connection; nobody at the Natanz facility is connected to the Internet. It's constantly drilled into everyone at the monthly security meetings that the Internet will let foreign spies in just as if you'd opened the gates of the facility and invited them in.

As you browse the lists of papers wondering which one to read first, a tiny program is copying itself, unknown to you,

I. Watson, *The Universal Machine,*
DOI 10.1007/978-3-642-28102-0_12,
© Springer-Verlag Berlin Heidelberg 2012

from the USB drive onto your computer. The laptop has a state-of-the-art anti-virus security program running, but it detects nothing. Later in the day you take your laptop onto the production floor of the facility and connect it via a cable to a sophisticated computer controlled centrifuge. This centrifuge, and the hundreds of others at Natanz, enriches uranium by spinning a gaseous form of uranium at unbelievable speeds, separating the lighter uranium-235 isotope from the heavier uranium-238 isotope.

The equipment you've connected to is made in Germany by *Siemens*; a SIMATIC WinCC/Step 7 controller. The small stealthy program that came from the USB drive is waiting for just this model of *programmable logic controller* (PLC). It exploits an unknown security hole in the PLC and changes a few bits of data. You are still completely unaware that anything is wrong. Over the next few days the infection, because that is how it acts, spreads from centrifuge to centrifuge and PLC to PLC. Now you start to notice that machines are reporting baffling errors, they're not operating at the correct speeds. One of your jobs is to correct this, so you make adjustments to the parameters of the PLCs. It doesn't seem to correct the problem in the way you expect. In fact, the machines no longer seem to be behaving logically at all. Then, disaster! Even though the centrifuges should be running well within their design tolerances their bearings are seizing, causing fatal damage to the equipment. Hundreds of expensive, but more importantly, virtually unobtainable and irreplaceable centrifuges have been damaged beyond any hope of repair. Natanz and hence the Iranian nuclear bomb program has been taken offline by the *Stuxnet* worm that was delivered on your USB drive.

So far this book has dealt with the many positive benefits of computers. However, just like any invention computers can be used for more sinister purposes. This chapter is about the dark side of computing.

THE BIRTH OF HACKING

The first people we might identify as hackers are actually our heroes from World War II; the Polish cryptologists Marian Rejewski, Henryk Zygalski and Jerzy Różycki who gave their secrets to Alan Turing and the other code breakers at Bletchley Park. If you visit Bletchley there is a memorial to the Poles commemorating their contribution to the Allied war effort. The words "*hacking*" and "*hacker*" were first used at MIT in Boston, not in the computing lab, as you might expect, but at a model railway.

The *MIT Tech Model Railroad Club* (TMRC) was founded in the 1940s and quickly grew into an enormously complex layout of tracks, trains and scenery. The signals, power and switching of the tracks was organized by *The Signals and Power Subcommittee* who designed all the circuits that made the model trains run. Initially they built a semi-automatic control system using telephone relays. This was replaced in 1970 by two PDP-11 minicomputers donated by the *Digital Equipment Corporation*. Yes, that's right they used sophisticated computers to control their toy railway. The words "*hack*" and "*hacker*" originated from TMRC members and meant a prank or jokester. Other words that entered the language of hackers and programmers such as: "*foo*," "*kludge*" and "*cruft*" also derived from the TMRC, but they're less well known outside computing.

Arguably the first recorded computer hack occurred at MIT in 1965 when a student, William Mathews, discovered a flaw in a text editor on the IBM 7094 that revealed the system's user password file to any user who was logged in. I don't know what he did with this information. There's also little recorded in the history books of hacking activity during the late 1960s and early 1970s. I think we can assume that curious and talented students would have been exploring their university computer systems and playing hacks on each other. Until computers were networked by the ARPANET in 1968 there wasn't much to explore though. You could hack into company mainframes if you had access via a remote teletype; but getting this access required using the telephone network.

CAPTAIN CRUNCH

John Draper, although in the US Air Force, was a bit of rebel. While stationed in Alaska in 1965 he helped his buddies make free long-distance phone calls by hacking the local phones. A couple of years later, still in the Air Force, he started a pirate radio station and after leaving in 1968 he operated another pirate radio station, in the San Francisco Bay Area, from his VW camper-van. He was by then a fully signed up member of the counter-culture. He still had an interest in phones though, later saying in an interview in *Esquire Magazine*:

> "...if I do it [phone phreaking], I do it for one reason and one reason only. I'm learning about a system. The phone company is a System. A computer is a System, do you understand? If I do what I do, it is only to explore a system. Computers, systems, that's my bag. The phone company is nothing but a computer."

He had learnt that a toy whistle that came free in boxes of *Cap'n Crunch* cereal made a perfect 2,600 Hz tone, the exact frequency that AT&T used to indicate that a trunk (long distance) line was free to route a new call. Using the whistle a phone phreaker could place calls without paying; thus, John Draper took the pseudonym *Captain Crunch*. Steve Wozniak's mother is said to have pointed out the *Esquire* article on phone phreaking to Woz who was fascinated by it.

John Draper aka Captain Crunch

Draper learnt, through the Silicon Valley grapevine that Woz wanted to meet him and one day he turned up at Woz's dorm room in Berkeley. He subsequently taught Woz and Jobs how to phreak and they went on to make and sell blue boxes, the electronic gadget that could generate the pure digital tones the phone network understood.

Phone phreakers met inside the telephone system itself by holding conference calls, a service usually only used by businessmen and government leaders. In these conference calls the phreakers would exchange information, tips and tricks and brag about their exploits. Often they would call foreign countries or route calls from a pay-phone all around the planet to the pay-phone next to them, just for fun. This was the birth of the hacker community. They were doing what hackers do, exploring a complex system by exploiting its loopholes and flaws and pulling pranks. Nobody was really harmed and so what if a big corporation lost a little money.

After the *Esquire* article phreaking became a national scandal; the telephone companies had lost face and had to act. Investigations were started, private detectives hired and Federal agents got involved. Draper was trailed and caught with a blue box in his car. He was arrested, convicted of "*Fraud by wire*," and sent to jail. Phreaking had become a dangerous activity and went even further underground. Many of the phreakers now transferred their passion to the emerging world of computers. Draper, like Woz and Jobs, became a member of the *Homebrew Computer Club* and later worked for Apple, writing the first word processor, called *EasyWriter*, for the Apple II. He still works in the computer industry and maintains a website where you can more information about his past exploits.[1]

CYBERPUNKS

Computer hacking entered the popular imagination with the 1984 movie *WarGames*, which told the story of David Lightman, a teenage boy who hacks his way into a Defense Department supercomputer inside a secure mountain bunker and almost starts a nuclear war. Although the movie was pure fiction several aspects of it were true. The idea that a kid could hack into computers owned by large corporations or the military was not fiction. Secretive underground clubs, with names like: *The Chaos Computer Club, Legion of Doom, 414* and *Warelords* were meeting on bulletin board systems (BBSes). Like the phreakers before them, hackers used pseudonyms like *Phiber Optik* and *Erik Bloodaxe*; they used the secret world of the BBSes to share their secrets and brag about their exploits.

---

[1] http://www.webcrunchers.com

Their basic technique worked as follows. Using any home computer they would use a 300-baud modem to connect by phone to a local university computer using a stolen or borrowed account name and password. Once logged in they could easily gain access to other computers on the new ARPANET and later the Internet. Hackers rarely hacked their local host since this access was vital to them. All they might do was set up some new accounts for their own use or to give to trusted friends. They didn't want to draw the attention of the local system administrators to the fact that their computer had been hacked in case passwords were changed and they lost their local access.

They tended to work at night, when system administrators were likely to be off work, so they could use the computers without their activity raising suspicion. They mostly would fit your stereotype of a hacker: young, almost exclusively male, curious and intelligent, though this was not always reflected by good school grades, socially inept and often with OCD tendencies. Their computing skills were self-taught and they were often addicted to hacking, working long hours through the night. Some, sadly, came from dysfunctional families and it seems as if the pure, cold logic of computers was comforting, peaceful and predictable for them—something they rarely found in their home lives.

Once inside their local host they could access other remote computers on the ARPANET; some of these were just other university computers, like their local host, which they might play pranks on. The hackers lived by a code of ethics: don't damage computers or data, copying data is fine but share what you find. Playing jokes on administrators was considered good sport.

When you log into a multiuser account computer there is usually a screen that welcomes you to the computer. This screen might just be a generic welcome message, but it often includes some news such as announcements of system maintenance. Hackers loved to leave their calling card, a short humorous message, on these screens so when users logged in they'd know the computer had been hacked.

A key prize was to gain *root access*. A multiuser account computer, typically running some variant of the UNIX operating system, is hierarchically organized. Normal users have the lowest level of access or *privileges*; usually being only able to read and write files to their own file-space, read from certain shared file-spaces and run certain programs. Above them might sit a user who had some control over a whole group of users; a professor with a class of students for example. They might be able to assign privileges for the whole class and write files to the classes' shared file-space. Above everyone is a *superuser* who has privileges for the entire system. This is typically the system administrator, or *sysadmin*, who has *root access*. "*Root*" is the conventional name in UNIX for the superuser who has total access and complete control of the system. If a hacker can *root* a system they have freedom to explore everywhere, look at and copy any data on the system, run any program they can find and install their own programs.

It might help you to think of these computers like a large building. The doors into the building are usually locked. A hacker gains entry by subterfuge, typically by stealing a key, a username and password, but sometimes by exploiting flaws in the building's security systems. Once inside the building some internal doors are open and some are locked. The hacker can explore the rooms behind open doors and look at their contents but needs new password keys to gain access to the locked rooms. If they can get root access then all the doors open, as though they had a master key. A few doors in the building may lead not to rooms, but to tunnels into other buildings. In this way a hacker can move around the network from computer to computer following what one hacker has called, "*the wild pleasure of exploration.*"

At the wedding in Rarotonga I mentioned in the last chapter, as a token of his trust the groom publicly gave his new wife a letter at the wedding dinner that allegedly contained his root passwords—she would now have superuser access to all his computers. Yes, it was a geeks' wedding!

It should be emphasized that hackers were mostly just curious and liked the intellectual challenge of overcoming a computer's security. Moreover, hacking wasn't illegal.

Computers were relatively new and the law hadn't caught up with the technology. If you break into a building you may be charged with trespass, possibly criminal damage if you break something. If you take anything, you'll be charged with theft. But if you break into a computer nothing physical has been transgressed; if you copy data the original still remains so what has been stolen? Moreover, you might be sitting in Australia hacking a computer in the US; you're not even in the same legal jurisdiction so what are they going to do, send the CIA after you?

As so often happens with teenagers, harmless pranks eventually get out of hand, particularly if they are trying to impress each other with the audacity of their exploits. The hackers of the *414* group became particularly infamous by hacking into dozens of high profile computer systems, including the *Los Alamos National Laboratory*, where the atom bomb had been developed, the *Sloan-Kettering Cancer Center*, where they accidentally deleted billing records, and *Security Pacific Bank*. Clearly there were potential issues for national security, the privacy of medical records and financial security. The FBI got involved and the 414s were eventually tracked down to Milwaukee. But, hacking wasn't illegal, so although they were made to stop and pay some restitution, they were not prosecuted. They did receive the notoriety they were perhaps after, featuring on a cover of *Newsweek* in 1983. Legislators also took notice and started the process of creating new laws to outlaw hacking.

1984 was an important year amongst the hacker community. Emmanuel Goldstein first published *2600: The Hacker Quarterly,* named after the 2,600 Hz tone the phone phreakers used. Actually *Goldstein* was a pseudonym of Eric Corley and was a reference to the hero of the underground movement in George Orwell's book *1984*. A "*Quarterly*" sounds a bit grand, it was really just a few pages stapled together and posted to anyone who sent in a stamped addressed envelope. *2,600* is still in publication, but now can be found on most good newsstands. 2,600 meetings were also quickly established all over the world, which still take place, even in my home country of New Zealand.

Another publication came out in 1984, William Gibson's debut novel *Neuromancer* that introduced the words "*cyberspace*" and "*cyberpunk*" into our language. Neuromancer was a big hit, becoming the first winner of all three of science-fiction's literary awards: the *Nebula Award*, the *Philip K. Dick Award*, and the *Hugo Award*. The story centers on Henry Case a talented computer hacker and a virtual reality data space called "*the Matrix.*" It's a great read and I imagine it was very popular with delegates at *The Chaos Communications Congress* the first, and now annual, hackers conference held in Hamburg, Germany, in the same year.

THE HACKER MANIFESTO

Hacking was an international affair, respecting no boundaries and hackers from different countries often met online in BBSes to share what they had learned. British hackers even went so far as to publish *The Hackers Handbook*. The underground hacker e-zine, *Phrack,* published a short essay called *The Conscience of a Hacker* by the hacker *The Mentor,* who had just been arrested. It's now better known as *The Hacker Manifesto* and it forms the ethical foundation for hacking.

DIGITAL UNDERWORLD

\/\The Conscience of a Hacker/\/

by

+++The Mentor+++

Written on January 8, 1986

=-=

Another one got caught today, it's all over the papers. "Teenager
Arrested in Computer Crime Scandal", "Hacker Arrested after Bank Tampering"...
        Damn kids. They're all alike.

        But did you, in your three-piece psychology and 1950's technobrain,
ever take a look behind the eyes of the hacker? Did you ever wonder what
made him tick, what forces shaped him, what may have molded him?
        I am a hacker, enter my world...
        Mine is a world that begins with school... I'm smarter than most of
the other kids, this crap they teach us bores me...
        Damn underachiever. They're all alike.

        I'm in junior high or high school. I've listened to teachers explain
for the fifteenth time how to reduce a fraction. I understand it. "No, Ms.
Smith, I didn't show my work. I did it in my head..."
        Damn kid. Probably copied it. They're all alike.

        I made a discovery today. I found a computer. Wait a second, this is
cool. It does what I want it to. If it makes a mistake, it's because I
screwed it up. Not because it doesn't like me...
                        Or feels threatened by me...
                        Or thinks I'm a smart ass...
                        Or doesn't like teaching and shouldn't be here...
        Damn kid. All he does is play games. They're all alike.

        And then it happened... a door opened to a world... rushing through
the phone line like heroin through an addict's veins, an electronic pulse is
sent out, a refuge from the day-to-day incompetencies is sought... a board is
found.
        "This is it... this is where I belong..."
        I know everyone here... even if I've never met them, never talked to
them, may never hear from them again... I know you all...
        Damn kid. Tying up the phone line again. They're all alike...

        You bet your ass we're all alike... we've been spoon-fed baby food at
school when we hungered for steak... the bits of meat that you did let slip
through were pre-chewed and tasteless. We've been dominated by sadists, or
ignored by the apathetic. The few that had something to teach found us willing pupils, but those few are like drops of water in the desert.

        This is our world now... the world of the electron and the switch, the
beauty of the baud. We make use of a service already existing without paying
for what could be dirt-cheap if it wasn't run by profiteering gluttons, and
you call us criminals. We explore... and you call us criminals. We seek
after knowledge... and you call us criminals. We exist without skin color,
without nationality, without religious bias... and you call us criminals.
You build atomic bombs, you wage wars, you murder, cheat, and lie to us
and try to make us believe it's for our own good, yet we're the criminals.

        Yes, I am a criminal. My crime is that of curiosity. My crime is
that of judging people by what they say and think, not what they look like.
My crime is that of outsmarting you, something that you will never forgive me
for.

        I am a hacker, and this is my manifesto. You may stop this individual,
but you can't stop us all... after all, we're all alike.

                        +++The Mentor+++

The Hacker Manifesto

269

The party was about to end; in 1986 the US Congress passed the *Computer Fraud and Abuse Act* that made hacking illegal. Most other countries around the world took note of this legislation and started drafting their own versions of the law.

Other people had also been paying attention to the exploits of the young hackers; people who were not motivated just by curiosity. In 1988 the *First National Bank of Chicago* revealed it had been the victim of a $70 million computer theft. Bank robbers no longer had to tunnel into the vault or hold up bank staff at gunpoint to steal money. They could rob a bank from the comfort of their own home using just a computer. Although First National was the first to reveal it had been robbed, it is widely believed that many other banks and credit card companies had lost money but never revealed their losses in order to maintain their reputations. Indeed, this practice still occurs today, since banks prefer their depositors to believe that their money is safe and secure.

Armed with the new law US Secret Service agents set about bringing the hackers to justice. *Operation Sundevil* was launched in 1990, resulting in dawn raids in 14 cities across the US. Hackers were arrested, their computers, modems, disks, and documentation were confiscated and they were brought to trial. Most of the hackers had never met before in person, only knowing each other by their pseudonyms in the BBSes. It seems that this is not the best way of fostering group loyalty and when investigators started pressuring the young hackers, several broke and gave evidence in return for immunity from prosecution.

Julian Assange

Similar raids and arrests were occurring in other countries around the world. The Australian federal police where the first to hack the hackers by remotely intercepting their data whilst they were hacking, leading to the arrest of several members of the Australian hacking group *The Realm*. The following year they arrested a young hacker in Melbourne, a member of *The International Subversives* known as *Mendax*. US Secret Services had put the Australian police onto him for accessing computers of the *USAF 7th Command Group*. Australian police also wanted him for hacking computers closer to home.

Mendax's real name is Julian Assange and his case is fairly typical. He'd written the rules for The International Subversives, amongst which were: "*Don't damage computer systems you break into (including crashing them); don't change the information in those systems (except for altering logs to cover your tracks); and share information.*" He pleaded guilty at his trial, but when it came to sentencing the judge was sympathetic, saying: "*there is just no evidence that there was anything other than sort of intelligent inquisitiveness and the pleasure of being able to – what's the expression – surf through these various computers.*" Assange was fined $2,000.

Such light sentences were fairly normal for genuine hackers and it seems that judges were able to distinguish the merely curious from the criminal. Unfortunately the police did not make this distinction. Many young hackers were arrested, but few real criminals were caught. In 1997 Assange provided research for an excellent book called, *Underground: Tales of Hacking, Madness and Obsession on the Electronic Frontier*. If you want to get the inside story on hacker culture in more depth I can highly recommend it. In true hacker spirit it is freely available online.[2]

WORMS, VIRUSES AND TROJANS

Running in parallel to our hackers was another development that has come to plague us. If we return to our metaphor of hacking as "*exploring a building*" you can appreciate that if a building is very large, it might take a very long time to fully explore. Hackers knew this and it frustrated them; a solution was to write small programs that could automate the exploration. They could set their code off to explore the building, perhaps searching for something specific, leave it to its task and return later to get the results. Hackers traded these pieces of code on the BBSes. An enhancement to these programs enabled the programs to copy themselves to other computers on the network if they could gain entry. They were able to *self-replicate* and are considered by some as a form of digital life.

Science and fiction were mirroring each other again. In 1975, John Brunner wrote a science-fiction novel called *The Shockwave Rider* in which a "*data-gathering worm*" travels through an "*electronic information web.*" Almost a decade later, Robert Morris, a Cornell computer science post-grad, released the *Morris worm* onto the Internet. The worm infected thousands of computers, often repeatedly, causing some to crash. In many other cases sysadmins noticed the infection and took their systems offline. The cost of disinfecting the computers was estimated between $10

---

[2] http://www.underground-book.net/

million and $100 million. Morris claims he just wanted the worm to be able to estimate the size of the Internet for him. The fact that he released the worm from MIT to hide his identity makes his claim questionable. Morris was the first person to be convicted under the 1986 *Computer Fraud and Abuse Act*; he was sentenced to three years probation, 400 hours of community service and fined $10,000.

The Morris worm caused its disruption just by increasing network traffic; it wasn't carrying a *payload*. Worms can carry small programs within them, a payload, that is designed to run on the host system. At first, in keeping with the hacker ethos, these payloads were jokes—one sent a picture of an owl with the phrase "*O RLY*" on it to the computer's printer. Worms increasingly became more dangerous though as their payloads became more malicious: the *ExploreZip* worm could delete files on an infected computer; some could send emails; others now install security holes (*backdoors*) in a system enabling hackers to take over the computer. These computers become *zombies*, slaves under remote control and like zombies they can be amassed into armies, called *botnets,* controlled by *botmasters* who typically use them to send spam or for denial of service attacks. You really don't want a worm to infect your computer.

You've almost certainly heard of computer viruses and your PC is probably running anti-virus software to protect it. It certainly should be if it's a Windows machine. Viruses, like worms, are also self-replicating programs, but viruses spread by attaching themselves to other programs or files. Email attachments and files on USB memory sticks are typical vectors viruses use to spread themselves. Computer worms in contrast are able to travel around the Internet independently. Once again the term "*virus*" was first used in a science-fiction story, written by David Gerrold and published in *Galaxy* magazine in 1969—hackers and computer scientists read a lot of science-fiction.

In 1971 Bob Thomas made the *Creeper* virus that infected DEC PDP-10 computers on the ARPANET. Its payload wrote the message "*I'm the creeper, catch me if you can*" on the screen. A program called the *Reaper* was created to find and delete the *Creeper*. A similar virus called the *Elk Cloner* infected early Apple computers and was spread by infected floppy disks. This method of infection meant that viruses could spread to computers that were not even connected to the Internet, something worms could not do. As with worms, the payloads of viruses have become increasingly malicious and preventing infection has become a big industry.

Another threat is the *trojan* named after the Trojan Horse, which you may recall from ancient mythology was presented as a gift and carried inside the walls of the besieged city of Troy. The wooden horse had Greek soldiers hidden inside it. When the Trojans were asleep the Greeks crept out of the horse, opened the city gates and the city was captured and burnt to the ground. Thus, a trojan is an innocent looking computer program, which may actually do something useful, that contains something dangerous inside it. Typically, like the Greeks at Troy, the payload of the trojan will open a door, a security hole, into the compromized computer. Trojans are not self-replicating and people who think they are getting something for free, like a game, download them from the Web without realizing. Trojans often install spyware that records all your online activity, including keystrokes so hackers can obtain your passwords and they can recruit your computer into a botnet. As you can see cyberspace is a dangerous place; you need to take care online and if you don't want to get infected use protection.

GOOD GUYS AND BAD GUYS

In a western movie you always know who the good guys and the bad guys are. By convention good guys wear white hats and bad guys black ones. This convention applies now to hackers. *White hat* hackers are the good guys; they are employed to penetrate a system's defenses and find security holes, known as *vulnerabilities*. Once a vulnerability is found they work with other white hats and computer hardware and software companies to *patch* the vulnerability before the bad guys find it.

Black hat hackers work covertly to *exploit* security vulnerabilities before the good guys find and patch them. If a black hat creates an exploit it may be used for the hacker's personal gain or glory, it may be secretly shared with other black hats or it may be sold to organized criminals. Cyberspace, as portrayed in the novel *Neuromancer*, is now a battleground between the forces of good and evil.

Some white hats try to infiltrate black-hat communities online, to find out where they are planning to attack. They also train people to be more secure; for example never leave passwords on post-it notes on your computer, you'd be surprised how common this is. Don't use the same password across different websites because if one is compromised all of them are. I've been told that security experts have also been known to test a company's defenses by leaving infected USB sticks lying around and hoping someone picks one up out of curiosity. It's been rumored that in 2008, US military computers were broken into after hackers scattered USBs across a car park at a US base in the Middle East.

There is a third type of hacker, the *grey hats*, who operate independently and without permission. If they find a vulnerability they may tell white hats or just publish an exploit in an attempt to force a rapid patch. Exploits are published online in places like the *Exploit Database*.[3] Because grey hats operate without authority they can end up being prosecuted for hacking. White hats and black hats often start their careers as grey hats turning to good, or to the dark side, as their nature leads them.

As computers become more vital to every aspect of our lives, both commercial and social, their security becomes more important. Computer security is now a major industry; universities teach courses on it and there are many publications and conferences devoted to hacking. Started in 1992 by the *Dark Tangent*, *DEF CON* is the most famous hacker conference. Its name is another hacker joke, a reference to the movie *WarGames* and the US Armed Forces

---

[3] http://www.exploit-db.com

"*defense readiness condition,*" or DEF CON. It could also be a reference to the letters "*def*" on the number "*3*" of a phone's keypad, followed by "*con*" from "*convention*" in honor of the phone phreakers—either way it's now a massive convention, held annually in Las Vegas.

Like the grey hat hackers themselves DEF CON walks a fine line between the legal and the illegal. Speakers and some attendees have been arrested, injunctions and restraining orders issued and various other legal threats made against the grey hats who, mostly, are trying to make cyberspace safer. Similar grey hat hacker conventions take place all over world. For example, in New Zealand *Kiwicon* 2011 had about 600 attendees from the security industry, government, the military, and the curious.

HACKTIVISM

"On the one hand information wants to be expensive, because it's so valuable. The right information in the right place just changes your life. On the other hand, information wants to be free, because the cost of getting it out is getting lower and lower all the time. So you have these two fighting against each other."

—Stuart Brand

Stuart Brand, the author of the *Whole Earth Catalog*, could have written this for today's cyberpunks. Their anarchist philosophy holds that information is being held prisoner by agents of "*the system*" who often demand money for its release. Hackers, crackers, phreakers, leakers and pirates have a duty to "*liberate*" information from its shackles. One hacker who subscribes to this worldview is Julian Assange.

After his prosecution for hacking Assange developed a variety of security related software applications before going to university in 2003. He doesn't seem to have been highly motivated because he barely managed to pass his courses in math, physics, philosophy and neuroscience. He dropped out in 2006 without graduating. As an ex-hacker

Assange still believed that making private information public empowers us, and he had the computing skills to create a global publication platform that lawyers and governments could not take down or silence—he called it *WikiLeaks*.

> *"To radically shift regime behavior we must think clearly and boldly for if we have learned anything, it is that regimes do not want to be changed. We must think beyond those who have gone before us and discover technological changes that embolden us with ways to act in which our forebears could not."*
>
> —Julian Assange

In his blog he writes: *the more secretive or unjust an organization is, the more leaks induce fear and paranoia in its leadership and planning coterie.... Since unjust systems, by their nature, induce opponents, and in many places barely have the upper hand, mass leaking leaves them exquisitely vulnerable to those who seek to replace them with more open forms of governance.*

WikiLeaks, established in 2006, operated under the radar for several years, publishing leaked documents from whistleblowers in investment banks, government departments and multinationals. A shocking video, shot from within a US helicopter, of the shooting of several unarmed Iraqi civilians, including a child and a Reuters journalist, was probably the highest exposure WikiLeaks received. That is, until Nov 2010 when WikiLeaks published 251,000 US diplomatic cables. The cables, many of them top secret, had been provided by Bradley Manning, a US Army soldier, who had copied them from a classified US security database onto a CD. Manning is alleged to have said, "*information should be free*" to a friend. The cables were credited with helping to fire-up the Arab Spring by highlighting the corruption of the regimes, and the contempt in which rulers held the populace.

The diplomatic cables, to the embarrassment of the US and many other countries are now free in the public domain. Manning however is not; he was arrested and charged with:

"...*communicating, transmitting and delivering national defense information to an unauthorized source and disclosing classified information concerning the national defense with reason to believe that the information could cause injury to the United States.*" The charges carry the death penalty!

In December 2010 *MasterCard, Visa, PayPal* and *Bank of America* suspended all transactions with WikiLeaks, making it very difficult for people to donate money to it. Assange claims WikiLeaks was losing over $80,000 a day in donations. A hacker group called *Anonymous* mounted a denial of service attack on the corporations' websites to express their support for WikiLeaks, who hadn't been found guilty of any crime.

A flag with the logo of Anonymous

Members of Anonymous are *hacktivists*, cyberpunks who use, "*illegal or legally ambiguous digital tools in pursuit of political ends.*" The earliest documented example of hacktivism was the WANK worm released in 1989 to protest against the nuclear arms race. The payload of the worm caused infected computers to show a protest message on their login screens.

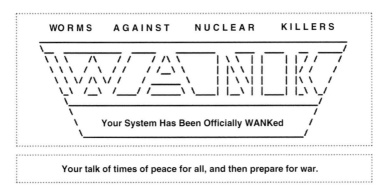

Screen shot of the WANK worm

This was a fairly harmless hacker prank, though hacktivists would go on to use more destructive methods. They use their skills to hide their online identities and penetrate the computer systems of their targets. Their attacks include: defacing the webpages of corporations and government agencies; denial of service attacks, a form of virtual sit-in that drives so much traffic to a website it can no longer handle it; email bombing where thousands of emails with large file attachments are sent to bring down mail servers; and logic bombs that wait inside a computer system for a certain date or trigger condition before activating their payload.

LulzSec is another group associated with hacktivism who have sometimes worked alongside Anonymous, though their motives are less clear. *"Lulz"* is derived from *"lol"* (remember the LOLCats) and *"sec"* from *"security;"* LulzSec aims to have fun by causing mayhem—they do their actions *"for the lulz."* In their manifesto it says *"we do things just because we find it entertaining...watching the results can be priceless."* They claim they like to draw attention to security holes and embarrass major corporates who should know better. In this respect they are acting like grey hats. Probably their most famous attack was in June 2011 when they compromised user accounts from Sony. They published the account names online because they believed it was negligent of Sony to store this sensitive data unencrypted.

LulzSec argue that by releasing lists of hacked usernames and informing the public of vulnerable websites, it gives people the opportunity to change names and passwords elsewhere that might otherwise have been easily exploited by black hats. They also believe businesses will be so worried that they might be publicly shamed they will consequently upgrade their security.

LulzSec loves to point out security flaws in organizations that should know better. For example, *Black & Berg Cybersecurity Consulting* recently issued a challenge to hackers, offering a $10,000 prize and a job to anyone who could hack their website and alter the homepage. Obviously, since they provide security advice to blue-chip companies this would be rather hard. They also sent a message to LulzSec's twitter account that said: *"@LulzSec Your Hacking = Clients for us. Thx."* The next day Black & Berg's homepage read: *"DONE, THAT WAS EASY. KEEP THE MONEY, WE DO IT FOR THE LULZ."*

CYBERWAR

In 2003 the *SQL Slammer* worm showed how truly dangerous Internet worms were becoming now we're all so interconnected. Exploiting a buffer overflow in Microsoft's SQL Server, used on thousands of webservers, Slammer infected 75,000 computers in just ten minutes. The irony was that Slammer exploited a known flaw in Microsoft's software for which a patch had been available for six months. Lazy sysadmins just hadn't bothered to patch their systems. The Slammer infection followed an exponential curve infecting 90% of all vulnerable machines within ten minutes. This caused a massive Internet wide denial of service and was a loud wake up call to the security world. A system-wide vulnerability could be exploited globally in just minutes before anyone had time to respond. The entire Internet and all systems that depend on it are potentially vulnerable to these attacks. The *Internet Storm Centre*[4] was established in 2000 to monitor the amount of malicious activity and it's geographic location on the Internet. It's a bit like a weather bureau that monitors the progress of hurricanes. Like the terrorist threat level it displays a *Green, Yellow, Orange* or *Red* alert levels along with detailed information on the nature of attacks.

In 2004 North Korea, one of the strangest authoritarian regimes on the planet, announced that it had formed a cyber-army that was busy hacking South Korea, the Japanese and their allies' computer systems. In 2007 a *spear phishing* incident in the office of the US Secretary of Defense led to the theft of classified material. It's not known where the classified material ended up, but the Chinese and Russians are suspected.

You may have heard of "*phishing*," it's the mailing of hundreds of thousands of emails by spammers in the hope that at least one recipient replies. The emails may claim you've won a competition, or that your bank wants you to change your password, or famously, that a Nigerian business-man wants your help moving millions of dollars out of his country. It's called *phishing* because the spammers are hoping someone takes the bait, as in actual "*fishing.*" *Spear phishing* is much more targeted, the *phisherman* contacts the target person, usually for some legitimate reason, gains their trust overtime and eventually tricks them into revealing some use-ful secrets, such as account names and passwords. Because of its personal nature spear phishing is very hard to combat, but it does require a significant investment of time from the attacker. Unfortunately, social networking sites like Facebook and in particular LinkedIn have made the job of the spear-phisher much easier. On LinkedIn you make your entire pro-fessional résumé available for the world to see. Anyone can see everywhere you've worked and everyone you've worked with if you accept them as a colleague. If a hacker assumes an ex-colleague's identity and refers to things that happened in

---

[4] http://isc.sans.org/

your professional life you're far more likely to open an email attachment from them—the attachment could contain a trojan or a virus!

In 2008 Chinese hackers claimed to have infiltrated the Pentagon and a couple of years later Google announced that its computer systems were coming under sustained and sophisticated attack from within China. It appears that cybercrime had become state sponsored. Just as political and liberation movements have received sponsorship from nation-states, now hackers are being encouraged by countries to work on their behalf.

It is alleged, though there is no hard proof, that the Stuxnet worm, that opened this chapter, was created either by Israeli military intelligence, or the CIA, or both. It was extremely sophisticated and served the objectives of Israel and the US by disrupting the Iranian nuclear program. Whether or not you approve or disapprove of the Iranian regime is immaterial; the danger is that Stuxnet opens up a whole new type of cyber-attack. Previously hackers just attacked computers. Stuxnet attacks the control systems of industrial equipment. A variant of Stuxnet could be created to bring down the electricity grid, turn off water treatment plants, and stop manufacturing equipment. A worm like Stuxnet can stop or harm anything that is in any way computer controlled, which is just about everything today.

Since last century we have become familiar during war to bombing raids that destroy airfields, communications systems, railways, bridges and factories. The first shots of the next war will probably take place in cyberspace; bringing down financial systems, electricity, and transport grids, and immobilizing factories without a single real bomb exploding. In response to the North Korean announcement the US in 2010 unveiled a new military force called *US Cyber Command* whose mission statement is:

"USCYBERCOM plans, coordinates, integrates, synchronizes and conducts activities to: direct the operations and defense of specified Department of Defense information networks and; prepare to, and when directed, conduct full spectrum

military cyberspace operations in order to enable actions in all domains, ensure US/Allied freedom of action in cyberspace and deny the same to our adversaries."

US Cyber Command emblem

The South Koreans, British, and Chinese have also announced the formation of similar cyber warfare units. In a worrying escalation the Pentagon has also announced that computer sabotage originating in another country could constitute an act of war and that the US reserved the right to respond to any such attack with traditional military force, including the nuclear option.

Hacking has travelled from the innocence of college pranks, via a cult of the curious, exploring the world's networks from their bedrooms, into organized crime. Along the way it has provided a new way for people to protest and for governments to spy and wage war. A hacker now could start a nuclear war by attacking a US defense computer—the movie *WarGames* was more prophetic that we thought. Universal machines may be leading us into a dangerous future.

Chapter 13

MACHINES OF LOVING GRACE

*I*t's 2030 and your personal assistant has just woken you from sleep. Your assistant isn't a person; it's a robot that analyzed your sleep patterns: eye movements, breathing, pulse and brain activity, to ensure you were only woken when you would feel most refreshed from your sleep. As you open your eyes, your bedroom ceiling flicks into life as a huge viewing screen. You can see the morning's news headlines, a weather report for the day and the view down your street, as if through a window—it's sunny. A local morning radio station starts playing. You scan the news headlines, and can read more details or watch video, just by wishing it to be so.

After a few minutes you get up and walk into the bathroom; the sound from the radio moves with you, though there are no visible speakers. As you enter the bathroom you start to think about your day ahead. A wall in the bathroom shimmers and your daily planner appears on it, along with your *To Do* list. You have a relatively clear day and you're going to see a local band in the evening with some friends. You think you'd like to hear the band's latest release and as you do, the radio station fades out and their latest hit starts playing. You take a shower and return to the bedroom, where your personal assistant has laid out your clothes. You dress and head for the kitchen, where your breakfast has been prepared: orange juice, cereal with fruit and yoghurt, and green tea.

As you're watching the morning news report the phone starts ringing; it's your mother and you answer. She appears as if standing in front of you and she reminds you about lunch on Sunday. You promise you'll be there on time; your father hates it if lunch is late and it's his birthday. You finish your breakfast and prepare to leave for work; as you leave you tell

I. Watson, *The Universal Machine,*
DOI 10.1007/978-3-642-28102-0_13,
© Springer-Verlag Berlin Heidelberg 2012

your assistant, *"I'll not be in for dinner, I'm going out this evening."* You're not sure why you do this, as it knows your diary as well as you do, but it seems polite, even though it's just a machine.

As you walk out the front door a small driverless vehicle stops outside; your ride to work has arrived. Nobody owns cars anymore; for a monthly subscription a vehicle is at your personal command whenever you want. You've rarely had to wait more than a few minutes for a ride; they seem to be able to anticipate your needs. Larger vehicles with more cargo space will arrive if you have to move bulky objects. For a small fee, the vehicles can even be used as part of a dating service. They will stop to pick up a person looking for a date whose profile matches, if the customer wishes. The *Lonely Hearts Club* rides have proven very popular.

The vehicle enters the motorway and its speed increases. Congestion and traffic jams are part of history, as the vehicles act like carriages in a train; moving quickly and safely, very close to each other. Road traffic accidents are almost unheard of; even *suicide by car* is very hard, as the vehicles are really good at avoiding pedestrians, including those who want to be hit. Your vehicle leaves the motorway, threads its way quickly into town and drops you outside your office. You get out, and then decide to go and shop for a birthday present for your father. He likes to fish, and you think some new lures might be appreciated. As you think this, an address for a fishing store appears in front of you, followed by directions: the store is just two blocks away. You think *fishing lures* and you see that they have a wide range in stock; so you set off towards the store. You don't have to be at work for any set time.

In 1967, Richard Brautigan, the *poet in residence* at Caltech in Pasadena, California, published a volume of poems called *All Watched Over by Machines of Loving Grace*. In the spirit of the times, 1,500 copies were given away for free. The title poem envisaged a future where computers did everything for people, freeing the human race from labor. This chapter is about how the universal machine will develop in the near future, and some of the changes this will bring about.

HOW FAR IS THE FUTURE?

I'll be honest: I'm approaching the writing of these last chapters with a little trepidation. I don't usually like futurologists who make their living predicting we'll all have personal jet packs, flying cars and computer chips implanted in us. It's always seemed to me to be just too easy. After all, you can make any prediction you like about 50 years into the future and nobody is going to check if you were correct. Also, futurologists like to pronounce on the big things, but always miss the small stuff that often really changes our day-to-day lives. For example, back in the 1960s futurologists were predicting that we'd be living on the moon, and tourists would stay in orbiting space hotels. Obviously we're not living on the moon; nobody has been there since the last Apollo mission in 1972. We don't have orbiting hotels, unless you count the *International Space Station,* but I suppose there have been a handful of space tourists; the multi-millionaires who've paid the Russians to take them up. So, totally wrong on the moon thing, mostly wrong on the space hotels and only partly right on the tourist thing. Not a very good record.

But what about the small stuff nobody predicted? For example, nobody predicted that by 2011 I wouldn't be able to remember the last time I went into a bank. My bank doesn't actually have physical branches; it's entirely online. I don't have a checkbook; all my transactions are done online. This might seem inconsequential, but it's saved me so much time and made a huge difference to the way I manage my finances. In comparison to my father I spend less time "*banking*" than he did, yet have far more control over my accounts than he ever dreamt of. To my mind, that's a profound effect brought about by computing.

Let's take another example: families used to scrimp and save so they could afford a set of encyclopedias that would be an invaluable educational resource for the entire family and the children in particular. Door-to-door salesmen actually used to sell encyclopedias, and you could take out a payment plan to spread the cost over several years—my family had a set. Today nobody buys them; everyone uses Wikipedia, and moreover we can access it from our mobile phones. This sort of change wasn't predicted either. So I'm naturally reticent to make grand predictions that will almost inevitably miss things that really do change our lives.

As with the other chapters in this book, I've done background reading and research on the subject, collected web resources and made lots of notes, but the process has also been rather different. All the previous chapters contain historical fact. For example, you can check out Woz's biography on Wikipedia or read his autobiography, *iWoz*, to see if I have the story straight. But for the future it's more a question of deciding which futurologists I want to read, and seeing to what degree they agree with one another and with my own intuitions. Wikipedia maintains a list (at the time of publication) of 118 futurologists living and dead.[1]

I thought it would be interesting to see if any of the people on this list have already appeared in this book—only four have: Douglas Engelbart, who envisaged augmenting human intellect; Grace Hopper, the pioneer of high-level computer languages; Vannevar Bush, the originator of the memex; and Stewart Brand, the publisher of the *Whole Earth Catalog*. The list does not claim to be complete, so I would add the following people from this book: Charles Babbage, for envisaging a machine that could be programmed; and Ada Lovelace, for seeing that symbol manipulation could be applied beyond math and algebra to music or poetry; Alan Turing, for much the same reasons, thereby giving us machine intelligence; Woz, for believing that computers could be personal; and Steve Jobs, for believing they could be user-friendly; Tim Berners-Lee, for seeing that people could publish and share information in a web; Mark Zuckerberg, for understanding the power of social networking; and William Gibson, for seeing that the world inside computers, *cyberspace*, is a place we can inhabit like the physical world. In a way, anyone involved with advancing computing is a

[1] http://en.wikipedia.org/wiki/List_of_futurologists

futurologist, since they have to envisage what the future will look like before they can make any advance.

Looking at Wikipedia's list, there are several names I recognized as futurologists who make predictions within the area of computing: Hans Moravec, Hugo de Garis, Jerry Fishenden, Joanne Pransky, Kevin Warwick, Michio Kaku and Ray Kurzweil. So I set about reading their blogs, and books, watching their numerous TV appearances and in one case, even a whole feature-length documentary. Some of what I read and saw was kind of obvious, some was interesting and some just plain daft.

I was particularly impressed by Michio Kaku's book *Physics of the Future: How Science Will Shape Human Destiny and Our Daily Lives by the Year 2100.* Kaku is a respected physics professor, and I think it was that his predictions were very well-grounded in science-fact, rather than science-fiction, which impressed me. He also structures his predictions in a sensible way: the *near future*—up to 2030; the *middle future*—up to 2075; and the *far future*—to 2100. The reason for this is that as you move from near to far, you can be less sure of predictions. Things in our near future are easy to predict, as they are already in development in research labs. Conversely, the far future is very hard to predict, since in the years between now and then we can't anticipate what scientific discoveries or inventions will utterly transform the world. For example, nobody was predicting the Web back in 1920.

MOORE'S LAW

The term *Moore's Law* was coined by Caltech professor Carver Mead in 1970 and named after Gordon Moore, the co-founder of *Intel Corporation,* who first described the underlying phenomenon in a paper called: "*Cramming more components onto integrated circuits*", published in *Electronics Magazine* in 1965. With the benefit of hindsight, which Moore didn't have, we can clearly see that the number of processors that can be placed inexpensively on an integrated circuit doubles every two years. Moore was the first to predict this.

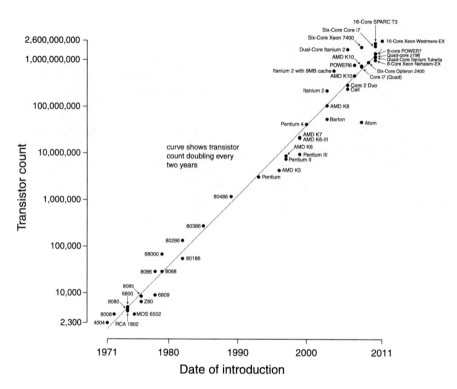

Microprocessor transistor count 1971–2011 & Moore's Law

The graph shown here has a logarithmic scale on the vertical "Y" axis and since the line drawn through the points representing individual processors is straight, this shows exponential growth. What this means in practical terms is that your smartphone has more processing power than all NASA had during the 1960s. In fact Moore's law can be extrapolated back in time as well; an Apple II personal computer had more processing power in 1977 than the entire Allies' computers at the end of World War II. Similar exponential growth has also been shown in hard disk capacity, the number of pixels per dollar on computer monitors and digital cameras, and the amount of network capacity on the Internet.

Put simply, all this combines to mean that our digital devices double in power, for the same price, every two years. This applies to everything electronic: computers, TVs, phones, etc.; if you put off purchasing now, a better model at the same price will soon appear. But if you wait longer, an even better model will appear at a cheaper price. So the wisest choice is to indefinitely put off any electronics purchase!

Although Moore's Law has profoundly affected the development of all the computers described in this book from ENIAC to the iPhone, I deliberately delayed writing about it until now. Whilst Moore's Law has been fabulous for the development of technology, ensuring faster and faster and more and more powerful computers, it's what happens when Moore's Law no longer applies that is crucial to our future.

In 2005, Moore himself predicted the demise of his own law: "*It can't continue forever. The nature of exponentials is that you push them out and eventually disaster happens.*" The laws of physics will trump Moore's Law by the end of this decade. Intel predicts this will occur at a 16-nm process node with 5-nm gates. When the wires and gates get really, really small—about five atoms thick—electrons begin to jump across the wires and short-circuit the chip. This is a physical behavior of electrons and makes it impossible to shrink transistors more. When transistors can no longer be made smaller, the only way to still double the transistor count every two years is to build upwards. If you look at the top right of the Moore's Law graph, you'll see that many of the processors have names like *dual-core*, *quad-core* and *six-core*. This tells us that these processors are made up respectively of two, four and six identical processors stacked together. Stacking silicon, though, has its own problems with heat dissipation and interconnections within the cores. So making larger, more complex silicon wafers and linking them together in a single package is only sustainable up to a point. The zenith of silicon-based integrated circuit design is finally visible on our horizon, as is clear from the recent proliferation of multi-core processors.

Futurologists, like Ray Kurzweil, hope that some new technology, like optical, molecular or quantum computers (which we'll discuss later) will come to the rescue of Moore's Law and enable us to continue to enjoy exponential growth in computing power far into the future. Whilst computer scientists, even within my own department, are exploring these avenues, their research is still in a very immature state, so I wouldn't hold your breath.

It's a common but entirely mistaken belief, and one that is often espoused by futurologists, that Moore's Law or *The Law of Accelerating Returns*, as Kurzweil calls it, applies to all technologies. This is patent nonsense, which we can observe all around us. Look at that iconic symbol of the twentieth Century, the motorcar: its development certainly hasn't followed Moore's Law. In 1908 a *Model T Ford* cost $850; in relative terms that's equivalent to about $20,000 in 2011.[2] Our cars are more comfortable, more reliable, safer, and come in more colors (customers were famously told they could have a Model T in any color they wanted, so long as it was black), but a 1908 Model T could get you around all day and cost about the same as a modern car.

Another icon, the Boeing 747 *Jumbo Jet,* first flew commercially in 1970; it carried about 450 passengers, had a range of just under 10,000 km and a cruising speed of Mach 0.84. The Boeing 787 *Dreamliner,* which first flew commercially in late 2011, carries 250 passengers 15,000 km at Mach 0.85. The Dreamliner is 20% more fuel-efficient, but this represents *incremental* rather than *exponential* improvement. I see no evidence of Kurzweil's *Law of Accelerating Returns* when it comes to these technological icons.

A final example shows us that sometimes we see virtually no progress at all, even for decades. Consider the typical US family kitchen of the mid-1970s. There isn't a single appliance we have today, over 30 years later, that wouldn't have been seen in a kitchen in the 1970s: the electric stove, toaster, microwave, refrigerator, deep-freeze, coffee maker, popcorn maker, sandwich press, juicer, food-mixer, etc. Yes, there have been stylistic changes, driven by fashion, but nothing significant on the technological front. So despite what futurologists like to tell us, technology progresses in fits and starts; the exponential growth of silicon chips is an exception, not the rule.

---

[2] http://www.measuringworth.com

Since Moore's Law is going to reach a real physical barrier in the foreseeable future I've decided to use Kaku's temporal divisions (near, middle and far) but reposition them in terms of Moore's Law. The *near future* is from now until Moore's Law ends sometime in the next decade. The *middle future* will be post-Moore's Law and the *far future* will be just that: the far, almost unimaginable future, unconstrained by our current physical and technological knowledge, and even the known laws of physics.

THE NEAR FUTURE

In the near future, Moore's Law still applies and computer chips are getting cheaper and cheaper. In my country there is a saying: "*as cheap as chips*", referring to something that is inexpensive and plentiful. In 1951, scientists led by Alan Turing programmed a Ferranti Mark 1 computer, a commercial version of the Manchester Baby that Alan Turing helped design, to play music, which the BBC recorded.[3] In 2011, you can buy novelty greeting cards that contain a chip that plays a tune when you open the card. The chips in these cards have more processing power than the computer in Manchester did in 1951. After a few days, these cards and the chips they contain will probably be thrown away.

As we move into the 2020s, computer chips will be disposable, costing cents, and they will be embedded everywhere; in clothes, in every household and work object, in the walls, floors and ceilings, outside in lampposts and pavements—everywhere. They will also be connected to the Internet in an "*Internet of things;*" we are entering a world of *ubiquitous computing*.

---

[3] You can listen to the recording here: http://news.bbc.co.uk/2/hi/technology/7458479.stm

UBIQUITOUS COMPUTING

We already have our feet on the threshold of ubiquitous computing; our smartphones enable us to access the Internet 24×7 and they are quite powerful computers, able to do processor-intensive tasks like editing video. But you have to remember to take your smartphone with you when you go out. In the near future, computers will be all around you all the time. Computer chips will be everywhere, and combined with ultra-high bandwidth Wi-Fi and cloud computing, will mean that you will be constantly immersed in an environment that provides all of your computing needs at all times.

Computing will become a utility that you assume is always available, like electricity is today. When you visit someone's house, you don't ask, "*do you have any electricity?*"—you just assume they will have electric lights and power sockets. It turns out that in a very few years, John McCarthy will finally have been proved right. You should remember him from chapter six; he was the head of the Stanford AI Lab in the late 1960s. He was an opponent of the personal computer revolution because, as a pioneer of time-sharing, he couldn't understand why people wanted their own dedicated computers. It was so wasteful, as for most of the day all that processing power would be sitting idle. He thought that through time-sharing, computer processing power should be purchased just like you buy electricity or water—you pay for what you consume.

At the end of the 1960s the ARPANET was brand new, Wi-Fi didn't exist and computers were only available on university campuses, in large corporations and government establishments. For Woz and his friends, having your own personal computer at home was intensely liberating and something to aspire to. Now thanks to wireless networking, cloud computing and cheap computer chips, we can compute anywhere.

CLOUD COMPUTING

It's probably a good idea to talk a little about cloud computing. Some of you will be familiar with it, but I'm sure for others it's a term you've heard but don't fully understand. If you already use Gmail, Hotmail or Yahoo mail, then your email is already in the Cloud. If you use a photo storing and sharing service like Picasa or Flickr, then your photographs are also in the cloud. You may also store other data with services like Google Docs, Dropbox or iCloud—all of these are cloud services.

You might imagine that Google, Apple and Amazon, who all provide cloud computing, must have an enormously powerful computer to deliver all their services; a computer perhaps like the new *Cray Titan*, which is rated at 20 petaflops; in English that's a quadrillion (1,000,000,000,000,000) mathematical calculations per second. But, actually you'd be mistaken; they use tens of thousands of relatively inexpensive PCs, linked together and housed in large air-conditioned ware-houses called *server farms*. The PCs in the farm are used to store data, but also to provide processing power for computa-tion through a *grid computer*. A grid is simply a network of inexpensive PCs connected together so all their processing power can be combined to create a very powerful virtual computer.

Many large companies and organizations now use the spare processing power of all the computers in their offices as a grid after office hours. So if you work in a large office, the PC on your desk may be processing the payroll, managing stock control or predicting the cost of oil in the future, all while you sleep. Amazon was a pioneer of cloud-based grid computing through its *Elastic Compute Cloud*. This allows people to buy computer processing as and when they need it.

Consider an event ticketing company: when tickets go on sale for a high-demand show, perhaps the reunion gig of a famous rock band, their website will be inundated with thousands of requests for tickets all within a few minutes. The ticketing company could buy all the expensive computing hardware they needed to safely handle this high volume of transactions, but for most of the time the hardware would be seriously under-used. This comes back to John McCarthy's point: why would the ticketing company want its own computers if they could rent the processing power when they needed it?

This is exactly what Amazon's *Elastic Compute Cloud* provides people. You can rapidly scale the number of processes you use as demand requires and then scale back when demand is low, only paying for the capacity that you actually use. You also reduce the infrastructure that has to surround computers: the technicians to maintain and upgrading the software, the security experts required to keep hackers out of your systems, and the engineers needed to fix or replace hardware when it breaks. Cloud computing treats computing, both data storage and processing, as a utility that you purchase, like electricity.

My wife is a primary school teacher; the children in her school are from 5 to 11 years old. Her school has just migrated its computer systems into the cloud. What does a primary school need with cloud computing, you might wonder; surely it's all paint and paper and scissors and glue at that age? No, the school had a computer server room because the teachers all use online systems to record children's grades and teachers' reports. Lesson plans and teaching resources are all stored on the servers, and the children themselves all have access to computers and have their own file storage. The school is really quite a technological place. A member of staff is in charge of managing the IT infrastructure, and an IT company provides technical and engineering support. The school recently migrated its entire computing infrastructure into the cloud so the school can now spend more time and money on actually teaching children. It also means the teachers and children can access their schoolwork from anywhere via the cloud.

COMPUTING AS A UTILITY

The cloud and the grid mean that we are moving into a future where all our data is stored in the cloud, along with most of our processing power. This future was first described by Mark Weisner while he was working at Xerox PARC around 1988. He called it *ubiquitous computing*; it's also known as *pervasive* or *ambient* computing. It represents a complete break from the desktop model of computing, as the machines fit seamlessly into our environment rather than the other way around. What this means is that wherever you go, providing you are connected wirelessly to the Internet, you will have access to all your data and all the processing power you may need on demand. This will have profound effects on the sort of hardware that we'll all be using. However, for once this doesn't mean more complex, harder to learn, or more expensive.

In fact, we can start to see this future happening in the very latest computers. Apple's ultra-thin MacBook Air is so thin not because of what it has, but because of what it *doesn't* have inside it. The MacBook Air doesn't have a CD or DVD/RW drive; Apple believes permanent local disk based-storage is a thing of the past. The Air doesn't need an Ethernet port for a physical network cable, since it relies on Wi-Fi. Its solid state hard drive uses less power than the old mechanical hard drives and is more reliable for a portable product. You see, as all your data is stored online you'll no longer need hundreds of gigabytes of storage on your computer; in fact you'll only need enough to cache your data before it is sent to the cloud, or as it is received from the cloud.

Since you'll be able to access cloud-based grid processing power when you want to edit that holiday video or play the latest game, you won't need more and more powerful chips and ever-increasing amounts of RAM in your computer either. You'll only need enough processing power to drive your video display, run the Wi-Fi communications and drive your web browser. Our mobile devices will also start to get simpler. Digital cameras are a good example; when you go on holiday it's important that you have enough memory in your camera to hold all the photographs that you'll take whilst you're away. For this reason digital cameras have removable memory cards; if you fill one card up, you can swap in another. If you run out of storage on your cards, you can easily buy extra cards whilst you're away. When you get home, you then have to upload all of your photos from all of your cards onto your computer, and then decide which photos to sync with any cloud photo-sharing services you use.

Soon digital cameras will need less storage, not more, and will never run out of memory. Just before I wrote this chapter, Apple released its new iPhone and an upgrade to its iOS operating system. The new operating system is fully integrated with Apple's cloud service called *iCloud*. One of the built-in features is *Photo Stream*, which uploads every photo you take with the iPhone's camera to iCloud. It also syncs new photos to every device you have registered with your iCloud account. Your photos appear on your home computer as if by magic, minutes after you've taken them from anywhere in the world.

This is the essence of ubiquitous computing: complex sequences of tasks being automated for you by processing in the cloud. As computers get smaller and smaller, they'll eventually become so small they'll virtually disappear; just as the chip that plays music inside a greeting card has. Academics are already experimenting with wearable computers. For example, the famous *MIT Media Lab* has a website dedicated to its work.[4] You can already buy a pair of Nike running shoes with a chip in them that can communicate with your iPod or smartphone to record your distance, your route and a host of other data. The neat thing about computing becoming ubiquitous is that it opens up a new range of ways that we will interact with computers and that they will interact with us.

One of the key elements of ubiquitous computing is that the computer is location-sensitive and always knows where you are. You may remember a scene from the Tom Cruise science-fiction movie *Minority Report*, in which he walks into a shopping mall. As he does so, a large screen greets him personally, and as he moves around the mall, advertising billboards change to show products and special offers that he might be interested in. This is not science-fiction. In the near future as you walk around town, you'll be made aware of special offers and discounts at businesses nearby. At the moment this can be done by sending messages to your GPS-enabled

---

[4] http://www.media.mit.edu/wearables

smartphone, but soon it will be directly to a heads-up display on your sunglasses or contact lenses.

Social networking services, like Facebook, will also be ubiquitous; as you look at a menu in a restaurant you may be told: *"your best friend liked #43, the stir-fried chicken with cashew nuts, last Tuesday"*. As more and more people are online 24/7, these recommendation services will become ubiquitous. Your car, which will also be connected to the ubiquitous computer system, will know where there are empty car parks near your destination, and when you get to the café, the staff will have your *"usual"* drink ready and waiting for you. Your car will have told them you're almost there, and the cloud will have reminded them what your *usual* is.

Many of these applications seem trivial, but in a way that's the point. Computing in the 1940s was about important, serious stuff: code breaking and designing atom bombs, then in the 1950s it was all about big business. In 2020, computers are ubiquitous and they sweat the small stuff, ensuring you get exactly the type of coffee you want when you want it—they are universal. Everything will be designed to make your life easier. Imagine a personal assistant working for you 24/7, knowing your habits, likes and dislikes, anticipating your every need and smoothing over any unforeseen problems.

HANDS-FREE

A key component of ubiquitous computing is the user interface. The computer chips will be woven into your clothes along with the Wi-Fi aerials. Your primary means of directing the computer will be by speech and thought—that is, if it hasn't already anticipated your needs. In 2011 Apple released an AI agent for the iPhone called *Siri*, which astounded everyone. We've seen voice recognition software before that can control our computers, but they have been very restricted to specific commands like: *"Open file"*, or *"Call Kate"*, spoken clearly and precisely. Siri is able to work in noisy environments and can handle imprecise conversational instructions like: *"cancel my meeting this evening and remind Kate we have dinner"*. How does Siri do this?

Siri is not a simple voice recognition system that works by linking a keyword like *"call"* to a specific action. Siri is the result of a very large research project funded by DARPA, the same agency that funded the development of the ARPANET back in the 1960s. The project, called CALO[5] for *"Cognitive Assistant that Learns and Organizes"*, was managed by SRI and employed almost 300 researchers from top university and commercial research institutions. Siri's goal was: *"building a new generation of cognitive assistants that can reason, learn from experience, be told what to do, explain what they are doing, reflect on their experience, and respond robustly to surprise"*. Apple bought Siri in 2010 because they want to make interacting with your digital devices as natural as possible. Remember: Apple brought us the mouse and the GUI because Steve Jobs didn't like the command line interface that computer programmers were used to. Siri represents a step change in how we interface with computers, and is essential for ubiquitous computing.

The other way we'll interact with our computers is by thought. This may sound crazy, but systems already exist that let you control computers just by the power of thought. The *Emotiv EPOC* headset looks a bit like a hands-free telephone headset, the sort receptionists and call-center employees wear, except it has 14 EEG electrodes that monitor your brain activity. The Emotiv costs $300, and after half an hour's training you can learn to move a ball on-screen in any direction you like, just by thinking about it. You can use the Emotiv for motion control in computer games, or the disabled can use it to control the cursor instead of using a mouse, or to steer a wheelchair. Controlling your computer by the power of thought is not science-fiction; devices similar to this will be miniaturized and in a decade or so will be commonplace. We will no longer control our computers with a keyboard and mouse, but by speech and thought.

---

[5] https://pal.sri.com/Plone/framework

AUGMENTED REALITY

The final key component of ubiquitous computing is how we receive information from the computer. In the far future this may be directly into our cortex, but in the meantime we know how this will work in the near future. Obviously computers will speak to us, as Apple's Siri does at the moment, but we will not be carrying around any device that has a computer screen. Instead we will receive visual images via special glasses and contact lenses or by projection directly onto our retinas.

Scientists at the University of Washington's *Human Interface Technology Lab*[6] are already developing contact lenses that can superimpose images into the wearer's field of vision to form a "*heads up display*." The research is still in its early days, but is already showing promise and has the potential to provide us with another feature of ubiquitous computing called *augmented reality*. In augmented reality, the computer projects information onto what we see. For example, in a shop it may perform a price comparison and inform us if the item is a good buy, or recommend we go to a different store where it's cheaper. In a museum it might show detailed information about the exhibit we are looking at, and in a cafe it might show us how many calories are in the chocolate brownie we want to order. If connected to a camera, concealed in glasses, jewelry or a pin, it might show us a magnified view of a distant object we're looking at. The reality we see is *augmented* with additional information generated by the ubiquitous computer.

Other researchers at the University of Washington are working on *virtual retinal displays* (VRDs). This comprises a tiny projector that can be mounted on glasses and projects three beams of red, green and blue light directly onto the retina of the eye. The wearer sees an image floating in space in front of them. A key advantage of VRDs is that because they project directly onto the retina, they can bypass any imperfections that the wearer has with their eyes, such as short or long sight or corneal problems.

---

[6] http://www.hitl.washington.edu/home/

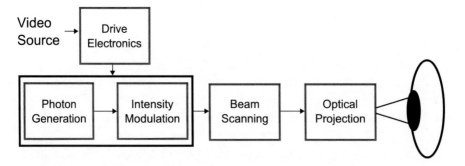

A diagram showing the workings of a virtual retinal display

At the moment VRDs are a little bulky, but in a few years time with the use of smart materials and clever optics, these should be able to be built directly into glasses that also contain headphones, a microphone, the thought control hardware and Wi-Fi connectivity. You'll hardly even notice you're wearing a computer at all. The irony is that as computers become ubiquitous, they also disappear.

A GRAND CHALLENGE

A rite of passage for many of us was learning to drive, and we can easily recall the feeling of liberation, tinged with just a little fear, when we first drove a car alone. In many countries, getting a driver's license is a moment that marks the transition from dependent child to independent adult. But in just a few years, people will not need to learn to drive, because cars will drive themselves.

Once again, a few of you might think I'm crazy, and that cars driving themselves is science-fiction and will never happen in our lifetimes. Well, you'd be wrong. As of 2010, Google has tested several cars that have driven autonomously over 140,000 miles (230,000 km) in an around San Francisco on normal streets at all times of the day. There was one accident: when a car driven by a person rear-ended the Google test vehicle that had correctly stopped for a red light.

It's not just Google researching driverless vehicles, but the US military and all the world's major car manufacturers. First, let's deal with the motivation: why do we want autonomous vehicles? The principal reason is safety. It turns out that people are really bad at driving (with the obvious exception of yourself). We get distracted and lose concentration, we drive too fast, we fall asleep, we break the rules, we're an accident waiting to happen. To put it into perspective, the US has lost over 6,000 troops in Afghanistan and Iraq since 2001. We see the tragic photos of the flag-draped coffins being returned home and watch grieving families at funerals, and rightly we all think this is shocking waste of lives. Last year 33,000 people lost their lives in road traffic accidents in the US alone. The year before the number was about the same, and the year before that. Next year another 30,000 or so will die, and that's not counting the hundreds of thousands each year who are seriously injured on the roads.

It's not that Americans are really bad drivers; in fact their road deaths per 100,000 of population are relatively low in comparison to countries like India, where the bloodshed is shocking. The United Nations estimated in 2004 that 1.2 million people were killed and a further 50 million injured on the world's roads. China recorded 96,000 deaths and India 105,000 traffic deaths. Traffic accidents are the leading cause of death from injury amongst children worldwide, and even in the US they are the sixth leading preventable cause of death. Clearly, if we can do something to stop this carnage, we should.

In 2004, DARPA announced the *Grand Challenge*, a long-distance competition for driverless cars. The US Congress authorized a prize of $1 million with the ultimate goal of making one-third of ground military forces autonomous by 2015. The military is also motivated by safety, but in their case they want to reduce the number of vulnerable troops put in danger delivering supplies and doing routine patrols.

Held on March 13, 2004 in the Mojave Desert, the course was a rough 150-mile route along roads, tracks and unmarked sections of open desert. None of the autonomous vehicles completed the challenge, with the winner travelling just 7.3 miles before getting stuck. Not a very promising start. In 2005, the teams were back for another Grand Challenge, and this time five of the teams completed the course, which was more mountainous and included running through *Beer Bottle Pass*, a winding mountain trail with a steep drop-off on one side and a sheer rock face on the other. One mistake would have been fatal, except of course there was nobody in the cars. A team from Stanford University won, driving a modified VW Passat.

Two years later in 2007, with the off-road driving now in the bag, the Grand Challenge became the *Urban Challenge*. The course involved 60 miles of urban driving; all traffic regulations had to be observed and the competitors had to deal with other traffic and obstacles, and merge into moving traffic. A team from Carnegie Mellon University and General Motors won the challenge, driving a modified Chevy Tahoe.

A car kitted out for the DARPA Urban Challenge

The basic technology that these autonomous vehicles are using is as follows. Firstly, they use global positioning satellites (GPS) to orientate themselves on a map, but GPS is not precise enough, particularly in built-up urban environments, to be used alone. A range of visioning systems is also used, such as radar, laser range finding, 3D laser modeling, and visual light imaging. My own university department has a group working with Mercedes on vision-based environment perception and driver assistance. The final component is software that combines data from the various sensing systems and the vehicle's systems, and decides how to steer, accelerate and brake appropriately.

At the moment all of these systems are bespoke and highly customized, but the aim of the car manufacturers is to make the equipment standardized and cheap, so it can be deployed across all their models in the near future. Of course, before driverless vehicles become common there will have to be legislative changes, since our laws require somebody to be in control of a vehicle.[7] Once we hand over that control, there will be several implications. Firstly, the terrible death toll will fall, hopefully to near zero, which will have a knock-on benefit to health care providers, since their workload will be greatly reduced.

Traffic cops will no longer be needed, since vehicles will be incapable of speeding, running a red light or parking illegally. Traffic jams will virtually disappear, as most jams are caused by accidents that won't happen. But even jams caused by "*sheer volume of traffic*" will be greatly reduced, as you won't see the *pulsing* that you currently experience in heavy traffic. The phenomenon of pulsing is why armies have always been trained to march in step. With one command, an entire column of men can move off and walk in unison, until directed to speed up or stop. If left to their own devices, people move at different speeds, and then some have to slow down, and

---

[7] The US state of Nevada passed legislation in 2011 that would allow autonomous cars on it highways by March 2012. An operator of an autonomous vehicle would still need a current state driver's license. The law defines an autonomous vehicle as "*a car that uses artificial intelligence, sensors and GPS to coordinate itself without active intervention by a human operator.*" The law also acknowledges that the driver does not need to be actively attentive if the car is driving itself. Without the law, inattentive driving would likely land you with some sort of reckless driving citation under the previous laws.

eventually even stop and restart. This causes the familiar pulsing we see in crowds, and also in heavy traffic. Driverless cars will be able to communicate wirelessly with each other to move in unison, as if a drill sergeant were in charge. Cars will also be able to form *road trains* and travel at high speeds with little distance between them, with the lead car nominally in charge of the entire train. This will mean that our roads will be able to carry much higher volumes of traffic than they do now.

These will be just a few of the changes I can guarantee you will see in the near future; but the story of the universal machine is not over yet. In the far future our lives may be changed beyond all recognition, and that is the subject of the final chapter.

Chapter 14

DIGITAL CONSCIOUSNESS

*I*t's February 16 2011 and you're sitting amongst the studio audience for the popular TV game show *Jeopardy*. Two of the show's previous champions are on stage: Ken Jennings, who had a 74 game winning streak that earned him over $2.5 million and Brad Rutter, the all time Jeopardy money winner, whose sharp mind earned him a staggering $3 million! In between them the third competitor isn't human; it's a computer called *Watson*. Where a person should be standing is a black screen with a globe shaped avatar on it that changes color as Watson thinks – green when it thinks it's correct, red when it's unsure.

Watson is relentless and quick, often leaving its human competitors no chance to get a toehold in the game. Watson is not always correct though; in the category of "*US Cities*" to the question, "*Its largest airport was named for a World War II hero; its second largest, for a World War II battle*." Watson answers, "*What is Toronto?*" That's totally wrong, Toronto's in Canada![1] Jennings and Rutter both correctly answer, "*What is Chicago.*"

By the final question Watson has an unassailable lead. Each of the contestants are given 30 seconds to write down their answer to: "*William Wilkinson's 'An Account of the Principalities of Wallachia and Moldavia' inspired this author's most famous novel.*" Rutter in last place reveals he has written, "*Who is Bram Stoker.*" That's correct. Jennings has

---

[1] To be fair to Watson there are also Toronto's in: Illinois, Indiana, Iowa, Kansas, Missouri, Ohio and South Dakota. The town of Tamo, Arkansas is also known as Toronto. Sometime it's wiser not to know too much.

I. Watson, *The Universal Machine*,
DOI 10.1007/978-3-642-28102-0_14,
© Springer-Verlag Berlin Heidelberg 2012

written: "*Who is Stoker? (I for one welcome our new computer overlords)*." The quizmaster and the audience laugh at the joke. Watson has written "*Who is Bram Stoker?*" and easily wins the competition with a total $77,147 to Jennings' $24,000. A computer has beaten the best humans in a competition that would seem to require considerable intelligence.

IBM developed Watson, named after IBM's founder Thomas Watson, just for a challenge. Watson is a whole series of computers; ten racks of IBM *POWER 750* servers with 15 terabytes of RAM and 2,880 processor cores operating at 80 teraflops (80 trillion operations per second). 200 million pages of information have been fed into Watson's memory including: books, movie scripts, song lyrics, poems and the full text of Wikipedia. It's not connected to the Internet but it can scan through its vast memory in less than three seconds. It has to understand the quizmaster's question, search its memory, infer several candidate answers, calculate the probability of each being correct, decide if it is confident enough to buzz in and then buzz in before its opponents.

First prize for the winner of the *IBM Jeopardy Challenge* is $1 million, second place $300,000 and third place $200,000. IBM said they'd donate the money to charity if Watson wins. We don't know what Watson thinks about this. We don't know if Watson thinks at all and this is the subject of the final chapter – will the universal machine think for itself and will we one day in the far future merge with them to become a new life-form?

THE IMITATION GAME

The quest for artificial intelligence (AI) goes back along way; we can trace the origins of automata to Greek mythology. *Talos* was a giant bronze man who protected Crete from invaders. In some versions of the myth the inventor Daedalus, who made wings for himself and his son Icarus so they could fly away from Crete, made Talos. Tales of fabulous human and animal automata abound in early histories from China, the Middle East and Europe. For example, the Islamic scholar

Al-Jasari describes programmable humanoids in his *"Book of Knowledge of Ingenious Mechanical Devices,"* published in 1206. These included a band of robotic musicians and a hand-washing automaton.

Leonardo da Vinci sketched a robot in 1495 and various mechanical automata were regularly displayed by the wealthy to entertain their guests. In 1769 a chess-playing machine, called the *Mechanical Turk*, created a sensation touring Europe, the US and the Caribbean astounding all who saw it. This machine played a very good game of chess, even defeating Napoleon Bonaparte and Benjamin Franklin. Inside the machine was a mass of gears and cogs; it looked terribly complicated. In 1820 a Londoner, Robert Willis, revealed the machine to be a hoax with a person cunningly hidden inside the mechanism.

An engraving of the Mechanical Turk showing its mechanism

Automatons wouldn't really play chess until the invention of the computer over a century later. The Mechanical Turk nicely leads us into thinking about the behavior we might inspect from intelligent machines. Ada Lovelace observed that the Analytical Engine was in theory capable of manipulating any symbols and not just numbers:

"Supposing for instance, that the fundamental relations of pitched sounds in the science of harmony and of musical composition were susceptible of such expression and adaptation, the Engine might compose elaborate and scientific pieces of music of any degree of complexity or extent."
–Ada Lovelace

A century later Alan Turing and others working on the first programmable computers had started to think about the implications of machine intelligence. Could a computer, by symbol manipulation, really compose a song or write a poem? In 1947 an American mathematician, Norbert Weiner, visited Turing in Manchester and they discussed machine intelligence. Weiner returned to the US and published a sensational book, called *Cybernetics,* in which he predicted the creation of electronic brains and entire factories of the future staffed by robots. Many scientists believed these ideas were nonsense, one of whom was the professor of neurosurgery at Manchester University, Geoffrey Jefferson. He used the occasion of being awarded the prestigious *Lister Medal* to attack the radical ideas originating from the Manchester Computer Laboratory. He said to a packed meeting during his *Lister Oration* on June 9 1948:

"Not until a machine can write a sonnet or compose a concerto because of thoughts and emotions felt, and not by the chance fall of symbols, could we agree that machine equals brain – that is, not write it but know that it had written it. No mechanism could feel (and not merely artificially signal, an easy contrivance) pleasure at its success, grief when its valves fuse, be warmed by flattery, be made miserable by its mistakes, be charmed by sex, be angry or miserable when it cannot get what it wants."

His oration was subsequently published in *The British Medical Journal* as *The Mind of Mechanical Man* and the *Times* of London paraphrased him by saying that unless the machine could, *"create concepts and find for itself suitable words in which to express them…it would be no cleverer than a parrot."* The debate was on: *"could machines be intelligent"* – it still rages today.

Turing set to work and published his seminal paper, *"Computing Machinery and Intelligence"* in 1950 in which he stated, *"I propose to consider the question, 'Can machines think?'"* He devised a test, now called the *Turing Test* for machine intelligence, which I briefly described in chapter four. The test is a form of imitation game based on a popular Victorian parlor game. In the parlor game a player has to guess if a person they cannot see is male or female just by asking them questions. The questions and answers are given on paper so voices don't betray the participants' sex. You win the game if you correctly guess the sex of your opponent and you loose if you're wrong.

Turing adapted this game for computers – could you tell if your opponent was a computer or a person? If not the computer has passed the Turing Test and must be intelligent. This neatly sidesteps Jefferson's insistence on the machine having feelings and emotions. Turing's test is explicitly about *imitating* intelligence; if the machine seems intelligent then it is. The *Loebner Prize in Artificial Intelligence*[2] is an annual Turing Test competition for *chat bots,* programs that let you have conversation with a computer.

So far no chat bot has won the Grand Loebner Prize of $100,000 and a gold medal, but each year a prize of $4,000 and a bronze medal is awarded to the best chat bot as voted by the judges. If you go online and try a conversation with one of these bots you will quite quickly come to the conclusion that they are not even close to being convincingly human. However, IBM's Watson is a whole different story. It doesn't engage in random conversation but it can understand natural language and correctly answer very challenging questions in seconds. Is it intelligent?

---

[2] http://www.loebner.net/Prizef/loebner-prize.html

THE CHINESE ROOM

The philosopher John Searle doesn't think Watson is intelligent. To his mind it's just manipulating symbols, there is no understanding within it. Searle agrees with Jefferson that intelligence is more than *"the chance fall of symbols,"* and he can prove it. In 1980 Searle, a professor at Berkeley, published a paper called *"Minds, Brains, and Programs"* in *Behavioral and Brain Sciences*. In it he describes a though experiment called the *Chinese Room*.

Imagine you are sitting in front of room with a locked door. You are a native Chinese speaker and you're able to ask questions of the room by writing them down in Chinese characters on paper, which you post through a slot in the door. The room replies in Chinese by sending written replies back through the slot. You spend time asking the Chinese Room factual questions and then progress to asking its opinions on the economy, politics the arts and you even ask it to tell you some jokes. All of its answers are perfect, just what you'd expect from an intelligent Chinese person. It has passed the Turing Test with flying colors – therefore it must be intelligent.

Then Searle unlocks the door and shows you inside the Chinese Room. Inside waiting to receive questions is an Englishman who doesn't speak or read Chinese. He takes each question and identifies the Chinese characters from a big ledger and writes down their corresponding numbers. He then, in a complex and laborious process, cross-references the numbers with other ledgers and indices and eventually obtains more numbers from which he obtains new Chinese characters that make up his answer. He writes these characters down and posts them back through the door.

Does the operator understand Chinese? The answer is obviously no. He's just following a laborious mechanical process, a program. In fact he needn't even know the characters make up a language called Chinese. Does the operator understand the questions? Again the answer must be no; he doesn't even know they are questions because he doesn't understand Chinese. Therefore the Chinese Room is not intelligent and never can be. It is, as Jefferson observed, just moving symbols around – there is no understanding.

AI researchers have struggled with Searle's simple thought experiment. It does seem to show that a universal machine by manipulating symbols can never be said to *think*. My take on this problem is to consider the example of flight. Do birds fly? Of course they do, not all birds, but most do and many fly very well. Do planes fly? Yes they do, but they fly in a very different way to birds. Planes don't have feathers and they don't flap their wings, but they can fly great distances and carry much more weight than even the largest bird. Therefore, flight is something that birds and planes both do but by using different methods. Birds are living animals that have evolved to fly and planes are engineered artifacts; machines that we have designed to fly.

Computers, like Watson, are machines that we have engineered to think. Watson isn't made of flesh and bone and it doesn't have a brain, but it appears to think, just not in the same way that we do. For some reason when it comes to intelligence and consciousness we are much more sensitive about the abilities of our creations. If we engineer a machine that performs as well as birds we proudly claim it flies, but if we engineer a machine that performs as well as or better than people in a game show we doubt it's thinking. I believe in the future the question of "*do computers think?*" will be one that most people will not even consider. We'll just all take it for granted that computers *behave* as if they are thinking and that's good enough. Philosophers will still be arguing about this in the future and the religious will always believe that machines can't have souls.

THE ROBOTS ARE COMING

"Robots are not people. They are mechanically more perfect than we are, they have an astounding intellectual capacity, but they have no soul."

– Karel Capek

Since the earliest days of computing we have worried that computers may one day take over the world and decide they don't really need us anymore. This fear seems irrational, but it is deeply rooted in the Frankenstein myth, which Mary Shelly so brilliantly described in her gothic novella of the same name. There is already one work place where robots have literally replaced people – factories. The word "*robot*" was first used in 1920, by the Czech writer Karel Capek, in a play called "*Rossum's Universal Robots*". The word "*robota*" in Czech means "*serf labor*" and carries with it a sense of drudgery. It's therefore a good word to describe machines that are intended to do repetitive manual work for us.

The first industrial robot, *Unimate*, worked on a *General Motors* assembly line in New Jersey in 1961. It did the difficult and dangerous job of moving castings and welding them to car bodies. The replacement of people by robots in dangerous environments is a common use for them. Robots can work tirelessly without a break, don't join trade unions and don't take holidays. So despite their expense they are increasingly replacing people in many factories. If you look at video from any modern car factory you'll see hundreds of industrial robots but only a few people. The Chinese electronics manufacturer *Foxconn*, which makes the Apple iPhone, announced plans in 2011 that it would dramatically increase the number of robots in its factories from ten thousand to one million by 2013. It currently employees over 900,000 people; it's not clear what effect its increasing use of robotics will have on its human workforce, but we can assume many will be made redundant.

In the previous chapter I wrote about the advent of driver-less cars and praised the benefits they will bring: safer roads, fewer traffic jams, the ability to do other tasks whilst travelling. But, spare a thought for the millions of people who earn their living driving: bus drivers, taxi drivers, delivery drivers, truck drivers – they could all loose their jobs. Thanks to Moore's Law, improvements in the mechanics of robotics and the ability to manufacture them cheaply we are close to a time in the near and middle future when almost all simple manual jobs will be done adequately by robots: street cleaning, stacking shelves, supermarket checkouts, burger flipping, the list is endless.

The Japanese, who are often eager to embrace new technology, already have a noodle bar where robots prepare your ramen noodles for you.[3] They ladle the broth, boil the noodles, put the noodles in the broth, sprinkle on toppings and serve. It takes them under two minutes to prepare a bowl and they make over 800 bowls a day! If they're not busy they entertain customers by performing tricks like spinning plates.

These robots are not the super intelligent robots we have come to expect from science-fiction, *Commander Data* from *Star Trek* or the *Terminator*, they are fairly dumb but are perfectly capable of performing mundane tasks. We do need to start thinking as citizens what happens to society if most manual workers lose their jobs. Henry Ford, the great pioneer of mass production, introduced a $5 per day wage in 1914, twice the usual pay at the time for Detroit. Ford explained his policy as "*profit-sharing,*" because it enabled Ford employees to afford the cars they were making. Unfortunately industrial robots don't have any income to spend. Who will buy the electronic gadgets Foxconn makes if nobody has a wage?

Another sector where we will shortly see many more robots is health care. Robots will be used in manual roles; cleaning and delivery around hospitals, but will also increasingly be used to provide health support. Once again the Japanese, along with the Koreans, are leading this research. Tasks the robots will initially perform will range from moving patients about; lifting them from beds and chairs; to health monitoring, taking blood pressure and other simple monitoring a diagnostic

---

[3] Watch video of the noodle robots http://youtu.be/5sVOSIUn7e0

tests. Large manufacturers such as *Toyota*, *Honda* and *Hyundai* are all investing heavily in health robotics viewing this as a major growth industry since both countries have a demographic problem – soon the elderly will outnumber the young and there will not be enough people to provide health care.

Health care robots will combine with ubiquitous computing to revolutionize how the elderly are looked after. An elderly person's home will be filled with cheap sensors that monitor how the person is sleeping, how often they turn during their sleep, and during the day move out of their favorite armchair – reluctance to move is often an early warning sign of illness. The fridge, pantry, cooker, refuse bin and toilet will all monitor food going in and out of the house or apartment. Changes in eating habits and excretions again may be an early sign of illness. A companion robot in the home would ensure that the elderly person correctly took any medications they needed and could act as a telepresence for healthcare professionals. Using the robot's senses they will see and hear what the robot sees and hears, and be able to conduct remote consultations with the elderly person. Those tragic stories you read about of an old person being found dead in their home when they had passed away days, or even weeks before, should become a part of history.

Professor Jim Warren, an expert in healthcare robotics in my department says: *"It's a crowning achievement of modern society in general, and health science in particular, that an ever-growing proportion of us can expect to reach advanced old age. But we can't have a healthcare worker in the home of every frail elderly person all the time. It is feasible, however, to have a robot present in each home. In our experience, the elderly (like everyone) naturally respond to the anthropomorphic character of even very basic robotic AI on a mobile platform as long as there's reasonable speech recognition. This opens up a wide range of helpful services that can allow people to live more independently for longer."*

Robots will also be used to assist the disabled with their mobility. The thought control devices I mention in the previous chapter can be used to control any device: a wheelchair, a computer, the TV... Exoskeletons are already being developed to enable people that can't move to walk and even run again;

it's feasible that with the assistance of ubiquitous computing and robotics many disabilities in the future will be no more disabling than short sight is to day. But, not all developments in robotics will be as beneficial.

GOVERNING LETHAL BEHAVIOR IN AUTONOMOUS ROBOTS

"Defense network computers. New… powerful… hooked into everything, trusted to run it all. They say it got smart, a new order of intelligence. Then it saw all people as a threat, not just the ones on the other side. Decided our fate in a microsecond: extermination."
– Kyle Reese from *The Terminator*, 1984

If you're a science-fiction movie fan you'll remember the exo-skeleton that Ripley in *Aliens* uses. It's like a forklift truck on steroids that you wear – it amplifies the operator's movements giving them superhuman strength. Similar, though more sophisticated, exoskeletons also feature in another Sigourney Weaver movie, *Avatar* – this time they are used as offensive systems for the infantry. As we've already seen, the military has invested heavily in computing and robotics: from SAGE, to the ARPANET, to DARPA's Grand Challenge for driverless vehicles. They have already researched exoskeletons for the infantry, enabling troops to carry more equipment for longer and to cover more ground faster.

Boston Dynamics are developing a robot for the US military called *BigDog*[4] that can act like a pack-mule. BigDog can walk and run over very rough terrain carrying heavy loads. It can go places that a wheeled or tracked vehicle cannot carrying an infantry troop's supplies and equipment. Boston Dynamics mission statement is: *"to build the most advanced robots on Earth, with remarkable mobility, agility, dexterity and speed."*

---

[4] Watch video of BigDog here: www.YouTube.com/BostonDynamics

A pair of BigDog robots

The battlefield of the future will contain fewer troops and more autonomous systems. Though strictly speaking not robots, unmanned aerial vehicles (UAVs) or *drones,* such as the *Predator* and *Reaper,* have already significantly changed the face of modern warfare. They are capable of fully autonomous flight, which means an operator can instruct them to fly from A to B or to patrol along a certain line or area; leaving the drone's operators to monitor its sensors. These include radar and a multi-spectral tracking system enabling the drone to see at all times of the day and in all weathers. They can stay in the air for over a day[5] and can carry four air-to-surface Hellfire missiles and two 500 lb. laser-guided bombs. Flying silently at 25,000 feet they have proved lethal in Afghanistan, North Western Pakistan, Iraq and presumably in top-secret operations not made public.

They are flown 24/7, if necessary, by shifts of operators in *Ground Control Stations* at bases like *Creech Air Force Base* near Las Vegas. Once a target has been identified the drone can lock onto it and track it if it's mobile. You've maybe seen video footage from a drone's cameras so you'll know how detailed the images can be. They claim to be able to recognize

---

[5] A Zephyr British drone has flown for over 82 hours nonstop.

faces and even to use automatic face recognition systems to identify suspects. Once a target has been acquired the officer in command will give the order to attack if the rules of engagement allow or send the information up the chain of command and await orders.

A Reaper drone in Afghanistan 2007

Although these drones can fly autonomously they do not acquire targets and fire missiles without human involvement. South Korea though has already announced it has deployed military robots along the demilitarized zone with North Korea that could act autonomously, firing a machine gun at an enemy it detects whilst on patrol. The title of this section is no joke; "*Governing Lethal Behavior in Autonomous Robots*" is a book by a colleague, Ron Arkin of Georgia Tech, which investigates the ethics of robots that are designed to kill.

Our worst nightmare, envisaged in the *Terminator* series of movies, may indeed be just over the horizon. Will people actually be hunted and killed by lethal robots in the future? It's certainly a possible future, which is why researchers in AI and Ethics are starting to seriously consider the implications of this technology. Unfortunately, if history is a guide then we should assume that if a technology can be used as a weapon it will be.

Since ancient times warfare has come with a degree of risk for both sides. When two warriors faced off against each other armed with swords each knew they could be killed or seriously injured. When a Reaper drone operator checks into work at Creech Air Force Base he or she expects to go home safely after the shift. People may have died in Afghanistan as a direct result of their day's work but they are under no threat themselves. This asymmetrical nature of *"warfare by remote control"* has led some to question its morality and even to excuse *"terrorists"* who bomb an American embassy because they feel it is one way they can retaliate against drone attacks. The increasing use of drones prompted The United Nations to appoint Philip Alston as a *Special Rapporteur* on extrajudicial, summary or arbitrary executions. Alston has said that the use of drones is *"not combat as much* as *targeted killing."* He has repeatedly tried to get the US to explain how they justify the use of drones to target and kill individuals under international law.[6]

RESISTANCE IS FUTILE

"The Terminator's an infiltration unit, part man, part machine. Underneath, it's a hyperalloy combat chassis – micro processor-controlled, fully armored. Very tough. But outside, it's living human tissue – flesh, skin, hair, blood, grown for the cyborgs."

– Kyle Reese, *The Terminator*

Several years ago I attended an AI conference in the US and went to a talk by Maurice Hurley. The talk was packed, standing room only; Hurley wasn't an academic or tech-entrepreneur, he was an elderly TV scriptwriter who had invented *the Borg* for the cult TV show *Star Trek*. He joked to the audience: *"I bet you're the only group I've ever spoken to who are rooting for the Borg!"* In case you're not a fan of *Star Trek* the Borg are cybernetic organisms, part machine and part humanoid, that have a communal *hive mind* whereby each Borg is connected

---

[6] Drone Wars UK is a blog that monitors the use of drones by the British and worldwide: http://dronewarsuk.wordpress.com/

to all the others. Their aim is to assimilate all other life forms in the universe into their *collective,* forming a perfect organism. The Borg has conquered thousands of systems in the *Delta Quadrant* and is now expanding into The Federation's part of space, where we live. In addition to assimilating all of the life forms they encounter into the collective they also acquire their technology and so are technologically far advanced of The Federation – resistance is futile!

The idea of part machine part human cyborgs has a long history; the *Six Million Dollar Man* and the *Terminator* to name two, but the reality is much more benign. There are already almost a quarter of a million cyborgs waking around with cochlear implants. These are surgically implanted electronic devices that enable profoundly deaf people to hear. Research is underway for an analogous device for the blind to directly stimulating the optical nerve or the brain itself. Other brain implants are being researched that may circumvent parts of the brain damaged by strokes or reduce symptoms caused by Parkinson's disease or clinical depression.

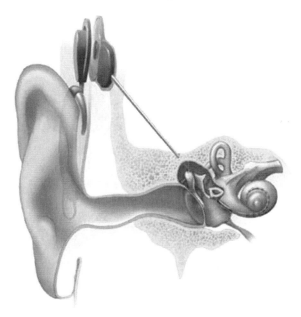

A cochlear implant

There are a whole host of ways we might use implants to improve our abilities in addition to ameliorating disabilities like deafness. Kevin Warwick, who is a professor of cybernetics at the University of Reading in England, famously experimented on himself by implanting an RFID (Radio Frequency Identification) chip in his hand in 1998.[7] This enabled him to remotely interact with sensors in his university building so doors would open for him and lights be switched on when he entered a room. Researchers at the University of Southern California's *Center for Neural Engineering* have demonstrated, on rat brains, that it is possible to implant chips into us that enable direct communication between computers and the brain. If we combine this with ubiquitous computing and the cloud it seems probable that many of us will choose to be part cyborg in the future.

SPOOKY ACTION

"If anybody says he can think about quantum physics without getting giddy, that only shows he has not understood the first thing about them."

– Niels Bohr

How will computers progress in the far future when Moore's Law no longer holds? One possible solution is *quantum computing*, but I warn you things quantum are very strange. We understand the *large* quite well now; thanks to Newton and Einstein we can predict the motion of the stars and planets and the effects of gravity, speed and time with stunning accuracy. But, the world of the very small, the sub-atomic quantum world, doesn't follow the same laws that govern the universe of the large. In the quantum world matter and energy, the stuff everything is made of, behave like particles and waves

---

[7] If you're not squeamish you can watch an RFID chip being inserted into a volunteers hand on YouTube http://youtu.be/HQ9-_soOToc

simultaneously and seem to be able to exist in two places at the same time.

A normal computer uses binary instructions of zeros and ones, whilst a quantum computer uses quantum bits, *qubits*, that also represent zero and one, and what's called a *superposition* of both (an amount of zero and one simultaneously). A normal computer can only store a single bit (zero or one) in a register at any one time, whereas a quantum computer can store more than a single bit. Adding qubits to a quantum computer increases the number of bits the computer can store and process exponentially. Remember from Moore's Law earlier that exponential growth in processing power is a very good thing.

However, there is a small problem. If you observe a qubit to find out if its value is zero, one or a superposition, your act of observation may make its value change. Erwin Schrödinger described this problem in 1935 as a thought experiment called *Schrödinger's Cat*. Imagine a cat sealed in a box with a flask of poison and a Geiger counter. The box is magically shielded from external quantum effects and so is only influenced by internal quantum effects. If the Geiger counter detects radiation, which it randomly may, the flask is broken releasing the poison and the cat dies. Since the cat's fate is decided by quantum effects, which we can't predict, this actually means that after a time the cat will be simultaneously *both* alive *and* dead! But, and here's the paradox, if you open the box the cat will *either* be alive *or* dead.[8]

Now, having a computer in which you can't observe results would be rather useless and here is where an even weirder quantum concept comes to the rescue. Quantum *entanglement* enables physicists to link or *correlate* two quantum objects together so that by observing one you can know the state of the entangled object. Quantum entanglements require no physical link and can occur between objects that are very far apart. Einstein called this, *"spooky action at a distance"* – I think he was using technical physics terms. It's as if we could observe Schrödinger's Cat by looking at its

---

[8] No actual cats were harmed in this physics experiment.

entangled cat on a different continent. Are you feeling giddy yet? As the famous physicist Richard Feynman said, "*anyone who claims to understand quantum physics doesn't understand quantum physics.*"

So if we entangle our qubits we can know the state of the entangled object by looking at its correlated fellow and we can therefore use our quantum computer. Research into building a quantum computer is still in its infancy. Photons have proved relatively easy to entangle, whilst electrons are somewhat harder. Optical systems have been made using photons by targeting lasers on beam splitters to generate qubits and super cooled superconductors have been used to create electron-based qubits. So far the largest quantum computer built has 128 qubits; developed by a Canadian company called D-Wave Systems Inc. they started marketing it in 2011 – it costs $10 million.

The D-Wave Systems quantum chip (photo provided by D-Wave Systems CC 3.0)

If we assume that the long awaited breakthrough into room temperature superconductors occurs, such as may be promised by the recently discovered new form of carbon called *graphene*, and that several other necessary breakthroughs take place then in the future quantum computers could provide us with unlimited processing power. Quantum computing will also have other important spin-offs because chemistry and nanotechnology fundamentally rely on understanding quantum systems. These are impossible to model and simulate accurately using normal computation; but quantum simulation will enable accurate modeling. This should enable great breakthroughs in materials science, which should further improve quantum computer chip design.

There is research into other unconventional methods of computing, which may also help crack Moore's Law. At the moment the top contenders are: *biocomputers* where DNA molecules or proteins are used to encode and process information; *molecular scale electronics* where molecules are used to build electronic components as small as a single molecule; and *photonic computing* where photons of light replace the electrons in our electronic devices to create photonic devices. Each of these research areas, and presumably others yet to be imagined, could result in computers of immense speed and power in the far future.

REVERSE ENGINEERING THE BRAIN

"It is the tension between creativity and skepticism that has produced the stunning and unexpected findings of science."
– Carl Sagan

In February 2008 the US *National Academy of Engineering* announced the *Grand Challenges for Engineering*[9] project and formed a committee of the wise and the visionary that contained amongst others, Google's co-founder Larry Page and futurologist Ray Kurzweil. The committee was tasked with identifying several grand engineering challenges to focus the attention of the world's politicians, scientists and engineers. Alongside worthy projects like providing universal access to clean drinking water and preventing nuclear

---

[9] http://www.engineeringchallenges.org/

325

terrorism, they highlighted the need to reverse-engineer the human brain.

The idea behind this is that we have failed to build an intelligent machine because we don't understand how the brain actually works. Therefore, in addition to the medical advances that may occur from a complete understanding of the brain we should be better able to design artificial intelligences. They're not looking for a purely physical description of the brain, though that is required, but a detailed functional description of what every synapse and nerve is doing at the lowest chemical and electrical level. The interrelationship of all the components would also be crucial to understanding things like memory formation, learning, recall, and emotions. This would be an enormous scientific undertaking, like *The Human Genome Project* of the 1990s, but they believe it's achievable. The aim would be to use this information to create computer chips that could mimic specific brain functions.

With $21 million in funding from DARPA, IBM is already combining principles from nanoscience, neuroscience and supercomputing in a cognitive computing initiative called *Systems of Neuromorphic Adaptive Plastic Scalable Electronics* (SyNAPSE). By reproducing the structure and architecture of the brain, the way its elements receive sensory input, connect to each other, adapt these connections, and transmit motor output, the SyNAPSE project[10] aims to emulate the brain's computing efficiency, size and power usage without being programmed. IBM's SyNAPSE learns how to perform tasks without following a program; it will behave like a brain, so perhaps one day we will create digital consciousness.

If we assume that at some time in the future we have successfully reverse engineered the human brain and can reverse engineer any particular human brain, because we're all slightly different, then this opens up some intriguing possibilities. Assume my brain has been reverse engineered and its accurate simulation is sitting inside a computer, does that make the

---

[10] http://www.ibm.com/smarterplanet/us/en/business_analytics/article/cognitive_computing.html

simulation me? Probably not, because the simulation doesn't have my memories and experience and isn't experiencing what I'm sensing at the moment.

So now assume that the reverse engineering process has discovered how memories are formed and stored so my life's experiences can be synced with the simulation. Now we potentially have a simulation that could be me. If we can hook it up to sensory systems, and that should be relatively easy, then we do have a complete simulation of me. It would also now be easy to augment the simulation with any computer program or information required: Wikipedia, a killer chess app, a mapping system, anything at all. My simulation would become a superhuman intellect.

It's still just a simulation though, the real me is still sitting here. The physical me, my body, will eventually die, but my simulation could continue forever. Computers can be repaired and copies or backups of the simulation could be made so it would never die. This of course opens up a Pandora's Box of ethical problems. If the simulations of me are conscious should they have legal rights, should they be able to be switched off or erased even though new simulations can be easily created? What will be the relationship between the real physical me and my simulations be, how many simulations could I have, what will be their relationship with each other? Supposing one simulation moves to a different place and has different experiences to another, then clearly the simulations are no longer the same, so are they then two unique individuals? I think you can see that this will keep the ethicists and philosophers very busy in the future.

Since the brain simulations are now just computer models it would also be possible to network different simulations and create a collective mind just like the Borg; a collective intelligence or *superbeing* would have been created. This network could also theoretically use quantum entanglement to transfer information instantly across the universe. We would have created a computer intelligence that could span the universe.

EPILOGUE

Our story of the universal machine is at its end; what a journey we've been on and what wonderful people we've met. From the mechanical engine envisaged, but never built, by Charles Babbage that sparked Ada Lovelace to first dream about machine intelligence, to quantum super intelligences that can instantly span the universe. We saw how the early pioneers of computing struggled to build calculating machines with electro-mechanical relays and how war caused a surge in innovation spurred by the need to process vast amounts of data. A genius, Alan Turing, proved that a single *universal machine* could solve any problem that could be represented symbolically. His break-through has been the guiding metaphor throughout this book because, if anyone can claim to have invented the computer, it's Alan Turing.

In the post-war years computer companies first used valves, then transistors and finally integrated circuits on silicon chips, which gave the name to a place crucial to the development of the computer. Silicon Valley in California was, and remains, the epicenter of computing. Stanford University, Berkeley, Xerox PARC, SAIL, and the high tech industries that surround them, created an explosion of innovation driven by the 1960s counter culture; reflected in Apple's slogan "*Think Different*." We've seen how brilliant kids, like Steve Wozniak and Bill Gates saw that they could build their own computers and software and a whole industry was born, led by a mercurial visionary – Steve Jobs.

People have a strong need to communicate and this led to computers being networked, first in ad hoc ways and finally by the Internet. Tim Berners-Lee, working in a physics laboratory, created a simple program called the WorldWideWeb, and everything changed. A second explosion of innovation occurred led by computer science students, again gravitating to Silicon Valley. Billions were made and lost during the dotcom bubble; but after the dust settled we were left with some great services and profitable companies: Google, Amazon, eBay, PayPal, etc. A third wave of innovation has recently taken place, once again driven by our need to communicate; social networking is pervading our culture and allows people from all over the world to share in each others' lives like never before. We truly are living in a global village.

But, all is not perfect; a group of young people, motivated by curiosity, the intellectual challenge and pranksterism have led the way into a dark and dangerous cyberworld. Criminals steal identities and money, and companies spy on each other; even more worryingly, nation-states are now planning to wage war in cyberspace. The military is even developing autonomous robots that have the potential to kill. Some of our nightmares that play out in fiction and on the screen are becoming possible.

Robots and artificial intelligence, ubiquitous computing and the cloud will transform society in the near future; yet we have no idea what to do with the hundreds of millions this revolution will make unemployed. Societies are in danger of dividing into the digitally enhanced and the disenfranchised. In the far future some predict we will all become qubits in various superpositions within simulations of reverse engineered brains. Computers and human minds will merge into a single super intellect. This intellect will be able to travel the universe at the superluminal speed of quantum entanglement; we will first become a galactic intelligence and ultimately a universal one, omnipotent and omnipresent – Gods.

We may exist forever in Turing's universal machine.

THE END

APPENDIX I

B elow is a complete transcript of British Prime Minister Gordon Brown's public apology to Alan Turing. The original is on display at Bletchley Park, England.

REMARKS OF THE PRIME MINISTER GORDON BROWN

10 September 2009

This has been a year of deep reflection – a chance for Britain, as a nation, to commemorate the profound debts we owe to those who came before. A unique combination of anniversaries and events have stirred in us that sense of pride and gratitude that characterise the British experience. Earlier this year, I stood with Presidents Sarkozy and Obama to honour the service and the sacrifice of the heroes who stormed the beaches of Normandy 65 years ago. And just last week, we marked the 70 years which have passed since the British government declared its willingness to take up arms against fascism and declared the outbreak of the Second World War.

So I am both pleased and proud that thanks to a coalition of computer scientists, historians and LGBT (lesbian, gay, bisexual and transgender) activists, we have this year a chance to mark and celebrate another contribution to Britain's fight against the darkness of dictatorship: that of code-breaker Alan Turing.

I. Watson, *The Universal Machine*,
DOI 10.1007/978-3-642-28102-0,
© Springer-Verlag Berlin Heidelberg 2012

Turing was a quite brilliant mathematician, most famous for his work on the German Enigma codes. It is no exaggeration to say that, without his outstanding contribution, the history of the Second World War could have been very different. He truly was one of those individuals we can point to whose unique contribution helped to turn the tide of war. The debt of gratitude he is owed makes it all the more horrifying, therefore, that he was treated so inhumanely.

In 1952, he was convicted of "gross indecency" – in effect, tried for being gay. His sentence – and he was faced with the miserable choice of this or prison – was chemical castration by a series of injections of female hormones. He took his own life just two years later.

Thousands of people have come together to demand justice for Alan Turing and recognition of the appalling way he was treated. While Turing was dealt with under the law of the time, and we can't put the clock back, his treatment was of course utterly unfair, and I am pleased to have the chance to say how deeply sorry I am and we all are for what happened to him. Alan and so many thousands of other gay men who were convicted, as he was convicted, under homophobic laws, were treated terribly. Over the years, millions more lived in fear of conviction. I am proud that those days are gone and that in the past 12 years this Government has done so much to make life fairer and more equal for our LGBT community. This recognition of Alan's status as one of Britain's most famous victims of homophobia is another step towards equality, and long overdue.

But even more than that, Alan deserves recognition for this contribution to humankind. For those of us born after 1945, into a Europe which is united, democratic and at peace, it is hard to imagine that our continent was once the theatre of mankind's darkest hour. It is difficult to believe that in living memory, people could become so consumed by hate – by anti-Semitism, by homophobia, by xenophobia and other murderous prejudices – that the gas chambers and crematoria became a piece of the European landscape as surely as the galleries and universities and concert halls which had marked out European civilisation for hundreds of years.

It is thanks to men and women who were totally committed to fighting fascism, people like Alan Turing, that the horrors of the Holocaust and of total war are part of Europe's history and not Europe's present. So on behalf of the British government, and all those who live freely thanks to Alan's work, I am very proud to say: we're sorry. You deserved so much better.

Gordn Brwn

APPENDIX II

WHERE ARE THEY NOW?

As I was finishing this book I realized that you might want to know what has happened subsequently to some of the characters I wrote about; after all since the history of the computer is quite recent many are still alive.

Charles Babbage – died at the age of 79 on October 18 1871. His son Henry Babbage built several working difference engines based on his father's designs in an attempt to rescue his reputation. Descendants founded the *Babbage Engineering Co.* in New Zealand.

Ada the Countess of Lovelace – died at the age of 36 on November 27 1852. The annual *British Computer Society* (BCS) conference for female undergraduates is called the *Lovelace Colloquium* in her honor. Since 1998 the BCS awards the *Lovelace Medal* to "*individuals who have made an outstanding contribution to the understanding or advancement of Computing.*"

Herman Hollerith – died at the age of 69 on November 17 1929.

James Ritty – continued to run saloons until he retired in 1895. He died at the age of 82 on March 29 1918.

Thomas Watson – became a prominent Democrat and a trustee of Columbia University and was active in the *Boy Scouts of America*. He died aged 82 on June 19 1956.

Howard Aiken – continued to work on new models of the Harvard Mark I up to the Harvard Mark IV. He was elected a

I. Watson, *The Universal Machine,*
DOI 10.1007/978-3-642-28102-0,
© Springer-Verlag Berlin Heidelberg 2012

Fellow of the American Academy of Arts and Science and in 1970 was awarded the IEEE's *Edison Medal* "*For a meritorious career of pioneering contributions to the development and application of large-scale digital computers and important contributions to education in the digital computer field.*" He died aged 73 on March 14 1973.

Grace Hopper – worked at the Harvard Computation Lab until 1949 and continued to serve in the US Navy Reserve. She retired with the rank of Rear Admiral in 1986 and worked as a consultant to *Digital Equipment Corporation* until she died on January 1 1992 aged 85. She was buried with full military honors in Arlington National Cemetery.

Donald Michie – continued to work in computing after WWII and became director of the University of Edinburgh's *Department of Machine Intelligence and Perception*. He later founded the *Turing Institute* in Glasgow and remained active in AI research into his 80s. He died in a car crash on July 7 2007 aged 83.

Bill Tutte – went to the University of Toronto, Canada after WWII and then the University of Waterloo where he stayed the rest of his career. In 2001 he was made an *Officer of the Order of Canada*. He died aged 84 on May 2 2002.

Tommy Flowers – continued to work for the *Post Office Research Station* after WWII and was involved in the development of ERNIE, a computer that generated the winning numbers for Premium Bonds (a type of lottery in the UK). He died aged 92 on October 28 1998. In 2010 an ICT center for young people, named the *Tommy Flowers Centre*, was opened in the London Borough of Tower Hamlets, where he was born.

Konrad Zuse – started several computer companies after WWII, the last of which was bought by Siemens. He died aged 85 on December 18 1995. The *Zuse Institute,* Berlin is named after him and in 2010 the centenary of his birth was celebrated by exhibitions, lectures and other events.

John Mauchly – was a founding member of the *Association for Computing Machinery* (ACM) and remained involved with computing for the rest of his life as a consultant. He received numerous awards and died aged 73 on January 8 1980.

J. Presper Eckert – became an executive in *Remington Rand* and remained there when it merged with *Burroughs* to become *Unisys*. He died aged 76 on June 3 1995. In 2002 he was inducted, posthumously, into the *National Inventors Hall of Fame.*

Maurice Wilkes – became a founder member of the *British Computer Society* and a *Fellow of the Royal Society*. In 1967 he was awarded the ACM *Turing Award*, computing's highest honor. He remained a professor at Cambridge University until he died aged 97 on November 29 2010.

Vannevar Bush – worked for various US government agencies after WWII and then served on the board of *American Telephone and Telegraph*; he was later chairman of *Merc & Co.* a pharmaceutical company. He was awarded the US *National Medal of Science* in 1963 and died aged 84 on June 28 1974.

Frederick Terman – became Provost of Stanford University and was a founding member of the US *National Academy of Engineering*. He was awarded numerous honors and died aged 82 on December 19 1982. The *Terman Middle School*, in Palo Alto, California is named after him.

Doug Engelbart – continued to work on his vision of augmenting human intellect as people moved away from the concept of networked collaborative computing towards the PC. His ideas fell out of fashion for over two decades and it wasn't until the 1990s that his ideas were once again recognized as revolutionary. In 1997 he was awarded the *ACM Turing Award* and in 2000 President Bill Clinton awarded him the *National Medal of Technology.* In 2001 he was awarded a *British Computer Society Lovelace Medal*. He serves on numerous advisory boards and is still an active computer science researcher.

John McCarthy – continued to work in AI and became an early advocate of using the Internet to foster free speech, virtual communities and social networks. In 1971 he was awarded the ACM *Turing Award* and subsequently received numerous other awards. He died aged 84 on October 24 2011.

Alan Kay – became *Atari*'s chief scientist after leaving Xerox PARC. He has subsequently worked with *Apple*, *Disney* and *Hewlett-Packard* and is a visiting professor at several universities. He is passionate about involving children in computing from as young as possible, and is closely involved with the *One Laptop per Child Program*. In 2003 he was awarded the ACM *Turing Award* amongst numerous other awards.

Steve Wozniak – remains an *Apple* employee but he has also been involved with numerous tech innovations and companies. In 2006 he founded *Acquicor Technology* with Gil Amelio, the ex-Apple CEO, and others to help develop new technology companies. He is known for his philanthropy, particularly in the area of education and computing. His annual *Wozzie Award* is presented to Bay Area high schools for innovation in the use of computers in business, art and music. He has been honored with numerous awards including the ACM *Grace Hopper Award* and a US *National Medal of Technology*. He is a keen Segway Polo player and claims to be a World Champion in Tetris.

Stewart Brand – writes and speaks and still explores ways of promoting his *Whole Earth Discipline*. He co-founded the *Global Business Network* in 1988 to promote global collaboration and the *Long Now Foundation* to encourage long-term strategic thinking.

Allan Alcorn – consulted and worked for several Silicon Valley start-ups after Atari and was an *Apple Fellow*. In 2000 *Lego* bought a games company he founded called *Zowie Intertainment*.

Dan Bricklin – founded, and remains with, a company called *Software Garden* that makes software tools including a program called *Dan Bricklin's Demo Program*, which allows you to demo programs before you create them. He was awarded an ACM *Grace Hopper Award* in 1981.

Bill Gates – is still on the board of *Microsoft* but stepped down as CEO in 2006 to dedicate himself to his philanthropic trust, the *Bill & Melinda Gates Foundation*. He has so far given over $28 billion to charity and expects to bequeath over 95% of his vast fortune. The foundation is the largest privately funded charitable organization in the world and has also received funds worth $30 billion from his friend Warren Buffet. The foundation operates in the area of health, poverty and education, in particular running vaccination programs in the third world. In March 2010 he lost his place as the world's richest man to Carlos Slim a Mexican telecommunications tycoon. I don't think Gates is too bothered.

Paul Allen – resigned from Microsoft in 2000. He has established a philanthropic trust called the *Paul G. Allen Family Foundation*, which operates in a wide variety of areas. He was a major investor in the online concert ticketing company *Ticketmaster* and is involved with commercial space flight projects. He lives a more flamboyant personal life than Gates, owning the *Portland Trail Blazers* NBA team, and the *Seattle Seahawks* NFL team. He also owns one of the world's largest private yachts, the 416 feet long *Octopus*, which carries two helicopters and two submarines.

Steve Ballmer – is CEO of Microsoft and the 46th richest person in the world according to *Forbes* magazine.

Tim Berners-Lee – founded the *World Wide Web Consortium* (W3C) in 1994 to ensure that the Web would be based on royalty-free technology. He has worked for the British government on digital government and is an advocate for *net neutrality*. He believes that access to the Internet is now a human right and that governments and other organizations should not restrict or monitor citizens' online activities. He was knighted by the Queen in 2004 and has received numerous other awards.

Jennifer Ringley – moved to Sacramento, California, worked as a web developer and for a social services agency. She has dropped out of public life.

Al Gore – failed to be elected US President in 2000 in a bitterly contested election that ended in a recount in Florida and a Supreme Court battle. Gore won the popular vote by half a million votes but George W. Bush was elected President. He subsequently became actively involved in the environmental movement and toured the world giving a PowerPoint presentation about global warming that was made into a documentary film called *An Inconvenient Truth*.

Ray Tomlinson – seems to live a quiet life since inventing email. He has been honored with several awards.

Ward Christensen – works for IBM and has received two *Dvorak Awards for Excellence in Telecommunications*.

Marc Andreessen – founded and invested in several social networking sites including *Ning*. He is on the board of *Facebook* and *eBay*, amongst other companies, and founded a Silicon Valley venture capital firm called *Andreessen Horowitz* who are a major investor in *Skype*.

Stephan Paternot – is now an author and film producer.

Larry Page – is CEO of Google and an investor in the electric car company *Tesla Motors*.

Sergey Brin – still works for Google and is involved with its charitable arm *Google.org*. He is actively involved in alternative energy and commercial space flight. *Forbes* magazine rank him and Larry page as the joint 24th richest men in the world.

Jeff Bezos – is chairman and CEO of *Amazon.com* and also has an interest in commercial space flight, founding a company called *Blue Origin* to help, *"anybody to go into space."* He was *Time* magazine's *Person of the Year* in 1999 and is a member of the controversial *Bilderberg Group*.

Pierre Omidyar – has established a philanthropic investment firm called the *Omidyar Network*. It helps organizations in the areas of micro-financing, government transparency and social media. According to *Forbes* magazine he is the 145th richest person in the world.

Shawn Fanning – has founded several start-ups including *Snocap*, *Rupture* and *Path*, all in the area of social media.

John Lasseter – is Chief Creative Officer at *Pixar* and the Principal Creative Advisor for *Walt Disney Imagineering*. He owns a working steam locomotive and is a NASCAR fan. He has won two *Academy Awards* and a *Special Achievement Award*.

Jonathan Ive – is Senior Vice President of Industrial Design at *Apple*. He was named *Designer of the Year* in 2003 and awarded a CBE in 2006. Amongst other honors, *Fortune* named him *the world's smartest designer* in 2010. In 2012 the Queen knighted him for "services to design and enterprise."

Mark Zuckerberg – is CEO of *Facebook* and will become the world's youngest billionaire when Facebook has its IPO in 2012. He has pledged to give 50% or more of his wealth to charity. He was named *Time* magazine's *Person of the Year* in

2010. He lives simply in Palo Alto with a small white dog called, *Beast*. He posts regularly on Facebook.

The Winklevoss twins – finished in 6th place at the 2008 Beijing Olympics. Cameron now publishes a blog called *Guest of a Guest* that talks about high society nightlife. It's not clear what Tyler does.

Sean Parker – became a partner of the *Founder Fund*, a Silicon Valley venture capital fund started by Peter Theil. He is a board member of the music streaming service *Spotify* and is an active philanthropist. He still lives flamboyantly and owns a $20 million Manhattan town house.

John Draper (aka Captain Crunch) – is still involved in the software industry and until 2010 was Chief Technical Officer for the media delivery company *En2go*.

Julian Assange – is currently in the UK fighting extradition to Sweden, where an arrest warrant has been issued concerning his sexual encounters with two women. He lost his case and the subsequent appeal and is now taking an appeal to the UK Supreme Court

Anonymous – may be "*legion*," but their members have been arrested in the US, Britain, Australia, Spain and Turkey.

LulzSec – has itself come under attack by hackers and in June 2011 a group called the *A-Team* posted a list of LulzSec members online. LulzSec then claimed to be disbanding.

Michio Kaku – is a professor of theoretical physics at the City University of New York, New York. He is a regular broadcaster and writer on popular science.

Gordon Moore – is Chairman Emeritus of *Intel Corporation*. In 2001 he and his wife donated $600 million to Caltech, Pasadena; it was the largest ever gift to a higher education institution. In 2007 they donated a further $200 million to Caltech, to construct the world's largest optical telescope. He enjoys fishing.

Watson – IBM's computer, has gone to work in healthcare. It plans to help physicians by synthesizing a patient's medical history along with all extant medical knowledge to suggest diagnoses and treatment options. It will not replace doctors but assist them.

John Searle – is still a professor of philosophy at the University of California, Berkeley.

Ray Kurzweil – swallows 150 supplement tablets, drinks eight to ten glasses of alkaline water and ten cups of green tea a day, in an attempt to *"reprogram his biochemistry."* He has signed up have his body cryogenically frozen should he die before technology has advanced to the point where his consciousness can be transferred to a computer. He believes medical advances will then enable him to brought back to life and that he will never truly die.

FURTHER READING

In writing this book I've obviously had to condense and paraphrase many much more detailed histories and descriptions. Hard choices had to made as to what had to be included and what was optional. A lot of interesting material and stories didn't make it into the book so I encourage you to enquire deeper. In my research for this book I have used the following general resources on most chapters:

Wikipedia http://en.wikipedia.org/ – the most marvelous source of reference material I can imagine. It doesn't matter how arcane the topic there will be a Wikipedia entry on it. I write using two monitors. The monitor on the left shows the chapter I'm working on whilst the monitor on the right always has Wikipedia open. I can read an entry, check my facts on one screen and write on the other.

The Computer History Museum http://www.computerhistory.org/ – a good place to double check Wikipedia and for in depth coverage of computer history. If you're in the vicinity of Mountain View, California you should go and visit them.

Reference sources used for individual chapters are given below.

I. Watson, *The Universal Machine,*
DOI 10.1007/978-3-642-28102-0,
© Springer-Verlag Berlin Heidelberg 2012

CHAPTER 2 – THE DAWN OF COMPUTING

The Cogwheel Brain: Charles Babbage and the Quest to Build the First Computer by Doran Swade. This is the definitive biography of Charles Babbage written by the man who led the project to rebuild his Difference Engine No. 2.

Ada, the Enchantress of Numbers: Prophet of the Computer Age by Betty Toole. Again the definitive biography about Lady Ada Lovelace, which tells her story as a narrative wrapped around her letters to others.

There are several websites devoted to Charles Babbage including:

> www.charlesbabbage.net
> ei.cs.vt.edu/~history/Babbage.html
> http://www.zyvex.com/nanotech/babbage.html

Websites about Ada Lovelace include:

> www.sdsc.edu/ScienceWomen/**lovelace**.html
> www.findingada.com

Two working replicas of Babbage's Difference Engine No. 2 have now been built. One is in the Science Museum, London and the other in the Computer History Museum in Mountain View, California. The Science Museum in London in addition has a collection of models and parts of Babbage's other machines, along with notebooks and other memorabilia. It also, rather strangely has half of Babbage's brain on display. I recently discovered that the other half is preserved at the *Hunterian Museum in the Royal College of Surgeons* in London.

CHAPTER 3 – MARVELOUS MACHINES

Making the World Work Better: The Ideas That Shaped a Century and a Company by Kevin Maney, Steve Hamm and Jeffrey O'Brien. This tells the history of IBM from Hollerith and the US Census through to the giant corporation that IBM has become. I used this book in this and subsequent chapters

Computer: A History Of The Information Machine by Martin Campbell-Kelly and William Aspray. The opening chapters of this book were a useful reference describing the development of early business machines and the formation of the first computer companies. The content of latter chapters is better described elsewhere.

IBM itself maintains a very comprehensive archive describing its history from Hollerith to the present day: http://www-03.ibm.com/ibm/history/

The Early Office Museum (http://www.officemuseum. com/) is a virtual museum that curates information about early office and business machines.

The LEO Computer Society (http://www.leo-computers. org.uk) maintains an archive of all things to do with the Lyons Electronic Office.

CHAPTER 4 – COMPUTERS GO TO WAR

Alan Turing: The Enigma by Andrew Hodges is the definitive biography on Alan Turing. Published in 1983 it was the first book that publicized Turing's great achievements and shabby treatment by the British establishment. I first read this when I was a graduate computer science student in the 1980s and it introduced me to Turing's genius. Though still considered the classic text it has been somewhat superseded by newer books that are able to reveal recently declassified information about Turing and his WWII code-breaking activities.

The Man Who Knew Too Much: Alan Turing and the invention of the computer by David Leavitt is a comprehensive biography covering all of Turing's life from childhood to his tragic suicide.

Turing – Pioneer of the Information Age by Jack Copeland, a leading expert on Turing, will be published in 2012 to mark the centenary of Alan Turing's birth. I'm sure like Copeland's other books it will be an invaluable reference.

The Essential Turing: Seminal Writings in Computing, Logic, Philosophy, Artificial Intelligence, and Artificial Life plus The Secrets of Enigma by Jack Copland republishes all of Turing's great scientific papers along side commentary.

The Ultra Secret: The Inside Story of Operation Ultra, Bletchley Park and Enigma by Frederick Winterbotham was the first book, in 1974, to break the secrecy surrounding Bletchley Park. Winterbotham was Chief of the Air Department of the Secret Intelligence Service during WWII, was based at Bletchley Park and reported to Winston Churchill. He was responsible for the organization, distribution and security of Ultra. That said, the book is inaccurate and scant when it comes to the codebreaking itself, but much better on how *Ultra* was used strategically across all theatres of the war.

Station X: The Codebreakers of Bletchley Park by Michael Smith (and other books by him) is another early book that breaks the secrecy surrounding Bletchley Park. It's stronger on the actual codebreaking than Winterbotham's book but newer publications are able to make use of more recently declassified material.

Codebreakers: The Inside Story of Bletchley Park edited by Sir FH Hinsley and Alan Strip is a fascinating book that collates the first hand recollections of many people who worked at Bletchley Park.

The Secret Life of Bletchley Park: The History of the Wartime Codebreaking Centre by the Men and Women Who Were There by Sinclair McKay is an interesting book that tells the stories of the men and women who worked at Bletchley Park during WWII. It is not a book about codebreaking but about the general life of people at Station X compiled from the memories of people, now in their 80s, who worked there.

Enigma: The Battle for the Code by Hugh Sebag-Montefiore is a dramatic and sensational retelling of the events, definitely a good read.

Colossus: The Secrets of Bletchley Park's Code-Breaking Computers edited by Jack Copeland tells the story, in detail, of the breaking of the German Lorenz machine code nickname "*Tunny*" by the Bletchley Park codebreakers and the development of Colossus the first electronic computer. Many people who were involved at the time, including Tommy Flowers the designer of Colossus contribute to this book.

Grace Hopper and the Invention of the Information Age by Kurt Beyer was used as a reference for this chapter and the following one.

Since Alan Turing is one of the greatest scientist who ever lived it is no surprise that he now has a considerable presence on the Web. Turing's alma mater *Kings College Cambridge* maintains the *Turing Digital Archive* (http://www.turingarchive.org/), which contains many of Turing's letters, talks, photographs and papers.

The academic and writer Jack Copeland maintains *The Turing Archive* (http://www.alanturing.net/) – the largest web collection of digital facsimiles of original documents by Turing and other pioneers of computing alongside articles about Turing and his work.

The writer Alan Hodges maintains *The Alan Turing Home Page* (http://www.turing.org.uk/turing/)

CHAPTER 5 – COMPUTERS AND BIG BUSINESS

From Airline Reservations to Sonic the Hedgehog: A History of the Software Industry by Martin Campbell-Kelly describes the early business uses of computers in the 1950s and 1960s well.

Computing in the Middle Ages: A View From the Trenches 1955–1983 by Severo Ornstein is an easy read and was used as a reference for this chapter and several latter ones.

Bright Boys: The Making of Information Technology by Tom Green is a useful book that I used as a reference source for this chapter and the preceding one.

CHAPTER 6 – DEADHEADS AND PROPELLER HEADS

What the Dormouse Said: How the Sixties Counterculture Shaped the Personal Computer Industry by John Markoff along with *From Counterculture to Cyberculture: Stewart Brand, the Whole Earth Network, and the Rise of Digital*

Utopianism by Fred Turner were invaluable in giving the background to how the counter culture influenced the development of the computer in Silicon Valley in the 1960s.

Dealers of Lightning: Xerox PARC and the Dawn of the Computer Age by Michale Hiltzik provides all the detail that you could possibly want about Xerox PARC and its remarkable inventions during this period.

http://www.netvalley.com and the *Santa Clara Valley Historical Association* (http://siliconvalleyhistorical.org) provide a good history of the development of Silicon Valley.

CHAPTER 7 – THE COMPUTER GETS PERSONAL

Accidental Empires: How the Boys of Silicon Valley Make Their Millions, Battle Foreign Competition, and Still Can't Get a Date by Robert X. Cringely is an entertaining history of the development of the PC and the software industry that grew up to support it. This book was the basis of the TV series *Triumph of the Nerds*. Robert Cringely was Apple employee #12.

Steve Wozniak's autobiography *iWoz: Computer Geek to Cult Icon: How I Invented the Personal Computer, Co-Founded Apple, and Had Fun Doing It* is really all you need to read about this remarkable man and the birth of Apple.

Hard Drive: Bill Gates and the Making of the Microsoft Empire by James Wallace and Jim Erickson provides a wealth of detail about the life and considerable achievements of Bill Gates.

CHAPTER 8 – WEAVING THE WEB

Casting the Net: From ARPANET to INTERNET and Beyond by Peter Salus gives a detailed history of the development of the Internet.

On the Way to the Web: The Secret History of the Internet and Its Founders by Michael Banks provides another history of the Internet's development. Its chapters on Bulleting Board Systems were particularly useful.

Inventing the Internet by Janet Abbate was another really useful resource and is perhaps more approachable than some of the other books.

Weaving the Web: The Original Design and Ultimate Destiny of the World Wide Web by Tim Berners-Lee obviously gives you the inside story of the invention and development of the Web. It's an entertaining and approachable book.

CHAPTER 9 – DOTCOM

Founders at Work: Stories of Startups' Early Days by Jessica Livingston tells the stories of many dotcom startups including: PayPal, Yahoo, Gmail, Flickr and many more. I didn't use much from this book, preferring to talk about the dotcom giants like Google and Amazon, but it was an interesting read.

Dot.con: How America Lost Its Mind and Money in the Internet Era by John Cassidy focuses on the successes and the failures of the dotcom bubble.

But, if you want to read more about some of the spectacular failures of the dotcom bubble then *F'd Companies: Spectacular Dot-com Flameouts* by Philip Kaplan is the book for you.

A Very Public Offering: A Rebel's Story of Business Excess, Success, and Reckoning by Stephan Paternot gives the inside story on the riches to rags epic that was theGlobe.com.

The E-Business Revolution & The New Economy: E-Conomics after the Dot-Com Crash by Gerald Adams gives a more economic rather than technological discussion of the times

In The Plex: How Google Thinks, Works, and Shapes Our Lives by Steven Levy is the new definitive story about all things Google; this is an excellent book.

One Click: Jeff Bezos and the Rise of Amazon.com by Richard Brandt will give you all the low-down on Amazon and its founder Jeff Bezos.

All the Rave: The Rise and Fall of Shawn Fanning's Napster by Joseph Menn is a thorough analysis of the brief rise and fall

CHAPTER 11 – WEB 2.0

Connected: The Surprising Power of Our Social Networks and How They Shape Our Lives – How Your Friends' Friends' Friends Affect Everything You Feel, Think, and Do by Nicholas Christakis and James Fowler may take the prize for the longest title but it was an interesting read. It takes a more sociological view than a technological one but it gives you the reasons why social networks have so much power.

The Accidental Billionaires: The Founding of Facebook: A Tale of Sex, Money, Genius and Betrayal by Ben Mezrich is a good read and it's the basis for the Oscar winning movie *The Social Network* – some of the scenes in the movie appear verbatim in this book.

The Facebook Effect: The Inside Story of the Company That Is Connecting the World by David Kirkpatrick gives you the detailed inside view of Facebook and mark Zuckerberg's world view – very interesting but perhaps a little too much detail.

Here Comes Everybody: How Change Happens When People Come Together by Clay Shirky shows how the power of social networks can be leveraged to change the world.

Wikinomics: How Mass Collaboration Changes Everything by Don Tapscott and Anthony Williams show how mass collaboration can change the way businesses, researchers, artists and political activists achieve results.

The Filter Bubble: What The Internet Is Hiding From You by Eli Pariser describes how the Internet has become a "*one way mirror*" reflecting your views, opinions and prejudices back to you.

CHAPTER 12 – THE DIGITAL UNDERWORLD

The Best of 2600: A Hacker Odyssey by Emmanuel Goldstein is a large collection of stories originally published in *2600 – The Hacker Quarterly*. If you want a complete history of hacking from the first phone phreakers to 2000 and beyond this is a great book.

the way we work, and shows how businesses and organizations must harness the new technology to succeed.

The Singularity is Near: When Humans Transcend Biology, by the famous futurologist Ray Kurzweil, is an optimistic and some would say fanciful vision of the future.

Infinite Reality: Avatars, Eternal Life, New Worlds, and the Dawn of the Virtual Revolution by Jim Blascovich and Jeremy Bailenson is in a similar vein but it lacks the quasi-religious undertones of Kurzweil's books

kurzweilai.net is Ray Kurzweil's website, where he maintains information in support of his futuristic ideas.

AAAI.org is the home of the *Association for the Advancement of Artificial Intelligence* and is the official home of serious AI academics. I attend its conferences regularly.

Stanford University is now offering its AI course online and for free. The course is taught by Professor Sebastian Thrun and Peter Norvig, Director of Google Research. If you have an interest in AI (and the time) I recommend you enroll and learn about how AI really works: https://www.ai-class.com/ You can watch the video content from the course without enrolling.

Printed by Publishers' Graphics LLC
AMZ20121115.11.35.66